Mollusks and Marine Environments of the Ten Thousand Islands

Mollusks and Marine Environments of the Ten Thousand Islands provides the first comprehensive overview of the shells and habitats that are present in the last unexplored coastal area of southwestern Florida. The mysterious and primordial Ten Thousand Islands, where the rivers and marshlands of the Everglades empty into the Gulf of Mexico, house a number of remarkable marine ecosystems, many shown here in detail for the first time. Primary among these are unique worm shell "reef systems," composed entirely of immense masses of vermetid gastropod mollusks. These previously unexplored and unstudied gastropod reefs, which are often many acres in size, are shown here to mimic coral reefs in their growth structure and represent the only large-scale molluscan reefs found anywhere on Earth. Living in association with the zonated gastropod reefs are a number of rare and unusual mollusks, some of which represent endemic species that are unique to the Ten Thousand Islands. These and many other southwestern Florida shells are illustrated throughout this book, along with detailed illustrations and descriptions of the marine and estuarine environments that dominate the archipelago and its adjacent lagoon systems.

Mollusks and Marine Environments of the Ten Thousand Islands

Edward J. Petuch
Department of Geosciences, Florida Atlantic University

and

David P. Berschauer
Museum Associate, Malacology Department,
Natural History Museum of Los Angeles County

CRC Press
Taylor & Francis Group
Boca Raton London New York

CRC Press is an imprint of the
Taylor & Francis Group, an **informa** business

First edition published 2022
by CRC Press
6000 Broken Sound Parkway NW, Suite 300, Boca Raton, FL 33487-2742

and by CRC Press
4 Park Square, Milton Park, Abingdon, Oxon, OX14 4RN

CRC Press is an imprint of Taylor & Francis Group, LLC

Library of Congress Cataloging-in-Publication Data

Names: Petuch, Edward J., author. | Berschauer, David P., author.
Title: Mollusks and marine environments of the Ten Thousand Islands /
Edward J. Petuch, David P. Berschauer.
Description: First edition. | Boca Raton, FL : CRC Press, 2023. | Includes bibliographical references and index. | Summary: "Mollusks and Marine Environments of the Ten Thousand Islands provides the first comprehensive overview of the shells and habitats that are present in the last unexplored coastal area of southwestern Florida. The mysterious and primordial Ten Thousand Islands, where the rivers and marshlands of the Everglades empty into the Gulf of Mexico, house a number of remarkable marine ecosystems, many shown here in detail for the first time. Primary among these are unique worm shell 'reef systems', composed entirely of immense masses of vermetid gastropod mollusks. These previously unexplored and unstudied gastropod reefs, which are often many acres in size, are shown here to mimic coral reefs in their growth structure and represent the only large-scale molluscan reefs found anywhere on Earth. Living in association with the zonated gastropod reefs are a number of rare and unusual mollusks, some of which represent endemic species that are unique to the Ten Thousand Islands. These and many other southwestern Florida shells are illustrated throughout this book, along with detailed illustrations and descriptions of the marine and estuarine environments that dominate the archipelago and its adjacent lagoon systems"-- Provided by publisher.
Identifiers: LCCN 2022011494 (print) | LCCN 2022011495 (ebook) | ISBN 9781032314792 (hbk) | ISBN 9781032314808 (pbk) | ISBN 9781003309949 (ebk)
Subjects: LCSH: Mollusks--Florida--Ten Thousand Islands. |
Mollusks--Ecology--Florida--Ten Thousand Islands.
Classification: LCC QL415.F6 P485 2023 (print) | LCC QL415.F6 (ebook) |
DDC 594.09759--dc23/eng/20220528
LC record available at https://lccn.loc.gov/2022011494
LC ebook record available at https://lccn.loc.gov/2022011495

ISBN: 978-1-032-31479-2 (hbk)
ISBN: 978-1-032-31480-8 (pbk)
ISBN: 978-1-003-30994-9 (ebk)

DOI: 10.1201/9781003309949

Typeset in Times Roman
by KnowledgeWorks Global Ltd.

MIX
Paper from responsible sources
FSC® C013056

Printed and bound in Great Britain by
TJ Books Limited, Padstow, Cornwall

Dedication

This book is dedicated to the following:

Linda J. Petuch, Eric and Rasa Petuch,

Brian Petuch and Kendra Berentsen, and Jennifer Petuch

and

Felicia Weisbrot Berschauer, Morgan, Jeremy and Lincoln Coker,

Jonathon, Tawni, Nora and Emmett Berschauer

and

A. Kenneth ("Kenny") Brown and Captain Craig Daniels

MOLLUSKS AND MARINE ENVIRONMENTS OF THE TEN THOUSAND ISLANDS

TABLE OF CONTENTS

ACKNOWLEDGMENTS

We thank the following the following for their invaluable assistance in the production of this book; without their help, this work would never have been completed. For helping with field logistics, loan of boats, and guidance throughout the islands, we thank: **A. Kenneth ("Kenny") Brown**, Chokoloskee Island, Florida; **Capt. Craig Daniels** and **Brandon Daniels**, Chokoloskee Island, Florida; **Jimmy Kidder**, Chokoloskee Island, Florida; and **Capt. Houston Brown**, Chokoloskee Island, Florida. For helping with data collecting and field exploration on the islands, we thank **Robert Owens**, Boca Raton, Florida; **Robert L. Eason, Sr.**, Paris, Tennessee; **Brian Ward**, Jupiter, Florida; **Rick Bell**, Westin, Florida; **Patrick Wilson**, Lake Worth, Florida; **Dave W. Fox**, Fort Myers, Florida; **Damien Schneider**, Athens, Ohio; and **Stephen Tressel**, Jupiter, Florida. For their help with tree snail identifications, we thank **Adrián González-Guillén**, St. Petersburg, Florida and **Frederick ("Pete") Krull**, St. Petersburg, Florida. For their kind donation of research specimens we thank **Robert Pace**, Miami, Florida; and **Donnie Benton**, Fort Myers, Florida. For the excellent bird and wildlife photography shown throughout this book, we give special thanks to **Michael Bruggeman**, Snellville, Georgia and **Morris A. Foster**, Jupiter, Florida. Their photographic skills truly captured the beauty of the Everglades and the Ten Thousand Islands and greatly enhanced this book. For technical assistance, manuscript review, and editorial guidance when assembling this book, we thank **Alice Oven**, **Michele Dimont**, and **Shikha Garg**, CRC Press, Taylor & Francis Group.

Dawn on Chokoloskee Island, Collier County, Florida, looking across Chokoloskee Bay. Photo courtesy of Morris A. Foster.

Shell Collecting in the Ten Thousand Islands: Responsibility, Ethics, and the Law

We have written this book to encourage interest in the very special marine environments that exist within the Ten Thousand Islands and the amazing lifeforms which inhabit them. We suspect that many of our readers will develop an interest in collecting some of the shells and marine organisms that are shown throughout this book. Along with this interest comes the responsibility of protecting these natural treasures and preserving their habitats. The following information is absolutely essential for anyone entering the Ten Thousand Islands area with the intent of collecting specimen sea shells. Since the Ten Thousand Islands archipelago is jurisdictionally divided into two distinct sections, the **Ten Thousand Islands National Wildlife Refuge** in the north (Collier County) and **Everglades National Park** in the south (Collier and Monroe Counties), two sets of laws and strictures have been established for the area. These are outlined here:

** **It is illegal to collect both living and dead shells within the Everglades National Park** boundaries. Dead beached shells are to remain untouched in order to preserve the ecological structure that is present in pristine beach ecosystems. **The Park's goal is to keep all natural systems the way they have always been**, going back to when the area formed in the Pleistocene. But photography of living animals and their resident ecosystems is welcomed and encouraged.

** **It is illegal to collect living shells in the Ten Thousand Islands National Wildlife Refuge in Collier County, but dead specimens washed up onto beaches can be collected.** Freshly dead shells that have washed up onto beaches can be picked up and collected without any legal problems. This type of beach shell collecting has become a major tourist attraction along the northern islands, especially Marco and Kice Islands and as far south as Camp Lulu Key.

** **Some specific areas of the Ten Thousand Islands National Wildlife Refuge are off limits** for any type of shell or marine life collecting, either live or dead. These areas, usually **bird sanctuaries**, are delineated by large signs and are often in the form of seasonal closures. The signs must be respected as they are notifications of sensitive bird breeding and nesting areas.

** If you intend to sell sea shells that were collected dead on the beaches of the Ten Thousand Islands National Wildlife Refuge, you must obtain a "License to Sell Salt Water Products" from the **Florida Fish and Wildlife Commission**.

The collection of voucher specimens as part of a field study has been a part of the scientific process for centuries and is just as relevant and important today. In compliance with local laws and regulations most of the specimen shells shown throughout this book were collected as dead specimens on beaches and in mangrove forests within the Ten Thousand Islands National Wildlife Refuge, primarily on Kice, Marco, and Cape Romano Islands and Camp Lulu, Round and Panther Keys. Others were live-collected decades ago by highly experienced shell collectors from Miami, before the present legal strictures were established and when the taking of live specimens was still legal throughout the entire Ten Thousand Islands area.

INTRODUCTION

The coastline of southwestern Florida, from Cape Romano in Collier County to Cape Sable in Monroe County, is characterized by extensive archipelagoes of small, jungle-covered islands and labyrinths of interconnected channels and lagoons. These islands and interconnected tidal streams, together with the impenetrable shoreline jungles, are referred to by geomorphologists and geographers as the **Reticulated Coastal Swamps** (White, 1970; Petuch and Roberts, 2007). This rather arcane designation for these beautiful island chains actually refers to the blotchy, mottled appearance of the islands and intervening lagoons when seen on aerial photographs of the region. This amalgam of small islands, channels, and lagoons is not a true swamp in the conventional meaning of the term, but is actually an expansive tropical tidal estuary, where salinities at any given locality vary with both the tidal cycle and with precipitation. Depending on the time of year, some of the nearshore lagoons become fresh water or brackish lakes, while at other times they become extensive pure salt water lakes or hypersaline pools. These wildly-fluctuating water conditions, sometimes occurring within the tidal cycle of a single day, have deleterious effects on many aquatic organisms and prevent both open oceanic marine species and pure fresh water species from becoming established within the Ten Thousand Islands area. The sources of the immense effluent of fresh water into the shallow marine environment are found in the multitude of streams, creeks, and sloughs that emanate from Everglades National Park to the north. The rainfall within the Everglades marshlands eventually finds its way to the Ten Thousand Islands through these channels.

Even though they appear to represent a contiguous and homogeneous array of heavily-forested archipelagoes, the Ten Thousand Islands are actually composed of two distinct types of islands. Studies of the underpinnings of these small landmasses, in transects running from the coastal lagoons (the "**Back Bays**" that are adjacent to the mainland) to the open Gulf waters, have shown that their foundations are completely different. For detailed information regarding the geological history of southern Florida and the Everglades see Petuch and Berschauer, 2021. Because of the geological differences described in the following sections, most of the Ten Thousand Islands can be classified as either belonging to the **Inner Ten Thousand Islands** or the **Outer Ten Thousand Islands** (Petuch and Roberts, 2007; Petuch and Myers, 2014: 156-168). These two main groups of islands grade together and form a blended continuum, with one type creating the stable base for the other. Very little geological or biological research has ever been undertaken within the Ten Thousand Islands, and the resident marine organisms are still largely unstudied and there has not been a large-scale ecological survey to date. This book is the first attempt at a close examination of the marine and estuarine environments and their associated organisms. As will be seen in the following chapters, several of the Ten Thousand Islands habitats contain ecosystems that are unique to the archipelago and these are described here for the first time.

Structure of the Inner Ten Thousand Islands and Back Bays

The Inner Ten Thousand Islands represent the closely-knit and compressed archipelagoes that have formed directly along the shoreline of Peninsular Florida. These are arranged around the large, lake-like brackish lagoons that parallel the seaward edge of the Everglades. Known as Back Bays, these prominent lagoons extend all along the coasts of Collier and northern Monroe Counties, and include Fakahatchee Bay, Chokoloskee Bay (the largest Back Bay), and Huston Bay in the northern section, and Chevelier Bay, Alligator Bay, First Bay, Big Lossmans Bay, Rodgers River Bay, and Ponce de Leon Bay in the southern section. The Back Bays, themselves, formed at the mouths of a myriad of rivers that flow directly from the vast Everglades wetlands. Some of the larger and more important of these Everglades drainage streams include the Ferguson River, Turner River, and Lopez River in the northern section of the islands and the Rodgers River, Broad River, Harney River, Lossmans River, and the Shark River in the southern section of the islands. These sloughs and meandering streams represent a transitional world between the freshwater wetlands of the Everglades sawgrass prairies and cypress forests and the marine world of the Gulf of Mexico.

Besides mangrove forests, the most prominent geomorphological features of the Inner Ten Thousand Islands are the large and extensive oyster banks that form within the Back Bays and tidal channels. These biogenic (formed by living organisms) landforms often cover immense areas and are composed of two types of oysters: the large Virginia Oyster (*Crassostrea virginica*), which prefers muddy, brackish water areas, and the smaller Crested Oyster (*Ostreola equestris*), which prefers more saline, or even hypersaline, quiet water areas. In many localities, such as Chokoloskee Bay, both the Virginia Oyster and the Crested Oyster occur together, often forming large reef-like biological structures that are referred to as "**bioherms**" (Figure 0.1). These ostreid bioherms are the only hard, solid structures on the floors of the muddy back bays and lagoons and they provide a base onto which Red Mangrove seedlings (**propagules**; see Chapter 1) can plant themselves and develop root systems. Once established, the pioneer Red Mangroves create conditions that are favorable for the colonization of the oyster bank islands by Black Mangroves. These two types of salt-tolerant trees eventually overwhelm and cover the underlying oysters, burying them in root mats, leaf litter, and accumulated mud and algae. The dead and buried oyster bioherms then form the bases for small mangrove islands which, in turn, grow together to produce larger islands.

As the mangrove islands develop on the oyster bioherms, the tidal currents, which change direction four times a day, carve out deep channels that run in labyrinthine fashion between and around the myriad of small landmasses. Over time, the smaller mangrove islands fuse together to form larger islands and the intervening smaller tidal channels fill in with sand and oyster shells, leaving only the largest and deepest channels to become permanent features. Because of tidal flow patterns, these larger drainage features generally are oriented in a north-south direction and form distinct aquatic barriers that separate individual groups of islands. Being filled with fish and other marine life, these large channels, referred to as **"Passes"**, often support large pods of Bottle-Nosed Dolphins (*Tursiops truncatus*), with each family group having its own distinct territory, and act as rookeries for predatory birds like the Osprey (*Pandion haliaetus*).

Structure of the Outer Ten Thousand Islands

The seaward edge of the Ten Thousand Island archipelago, which runs parallel to the inner archipelagoes and Back Bay lagoons, encompasses a set of islands that have a completely different geological foundation. Unlike the inner islands, with their oyster bioherms and muddy back bay environments, these islands are dominated by white quartz sand beaches, large sand-filled lagoons, and open oceanic sea water conditions. Referred to as the **Outer Ten Thousand Islands**, this parallel archipelago faces directly into the open South Florida Bight of the Gulf of Mexico and is subject to a **Longshore Current** along western Florida. This shoreline current, which is produced by the translation of forces during wave action, carries massive amounts of quartz sand down the coast from northwestern and western Florida. This pure white "Sugar Sand" is derived from glacially-eroded granite rocks in Canada and was later carried by streams into the main Mississippi River channel. Ultimately, the sand made its way down the river and into the Gulf of Mexico, where it came under the influence of the longshore current as soon as it was deposited in the ocean. This southward-flowing river of clean white sand runs along the shorelines of Louisiana and the Mississippi River Delta all the way south to Cape Sable ("Sand Cape" in French). There, it dissipates and spreads out into Florida Bay as a wide sediment fan that is buried beneath carbonate sediments before it reaches the Florida Keys.

Unlike the Inner Ten Thousand Islands, which formed on a base of oyster shells (buried oyster bioherms), the Outer Ten Thousand Islands formed in a far more exotic way, growing on top of massive biogenic structures composed entirely of worm snails (genus *Petaloconchus* in the gastropod family Vermetidae; see Chapter 3 of this book). These extensive "reef systems", here referred to as **Vermetoherms**, are the only large-scale Floridian biological structures known to be composed entirely of gastropod mollusks. The original worm shell aggregations grew on top of sea shell rubble piles in shallow intertidal areas with higher, more open oceanic salinities. The scattered smaller aggregations, some which began growing during the early Holocene Age (11,700 years ago), eventually fused together to form larger reef-like structures (Shier, 1969). These immense intertwined vermetoherms, some of which are now many acres in size, acted as sediment traps and produced environments that were favorable to mangrove tree colonization. Once mangrove trees became established on the vermetoherms, their prop roots (see Chapter 1) accumulated debris such as crushed oyster shells, sponges, algae, and decomposing mangrove leaves (mangrove peat), and produced soils that could support larger trees and salt-tolerant land plants. After a vegetated island had formed on top of the dead worm snail reef, it became a blocking area for the longshore current, allowing quartz sand to accumulate and creating broad white sand beaches. Often, these beaches, along with the adjacent sand flats, completely covered the vermetoherms, obscuring the original foundation of the island. These patterns of succession demonstrate that the Inner and Outer Ten Thousand Islands, although geologically different, are inextricably interconnected and form a continuum; intergrading from muddy brackish water areas with massive oyster banks to open oceanic salt water conditions that structures composed of worm snails.

In addition to the worm snail reefs and broad white sand beaches, the Outer Ten Thousand Islands also house a treasure trove of strikingly beautiful marine habitats that include biologically-diverse sponge "reefs" (bioherms), lush sea grass meadows, and extensive clam beds (bivalve mollusks), along with terrestrial environments such as vibrant tropical forests, fresh water springs and pools, and beaches covered with dazzling sea shells. On some islands, such as Pavilion Key, immense accumulations of the predatory Left-Handed Whelk (*Sinistrofulgur sinistrum*) and its principal prey items, the Campeche and Brown's Venus Clams (*Mercenaria campechiensis* and *Mercenaria browni*), cover large expanses of the beach areas (Figure 0.8). These extensive piles of large mollusks shells serve as mute testament to repeated mass die-offs of the local whelk and Venus Clam populations due to precipitous drops in salinity during the rainy summer months. The deposition of the shell piles along the beaches took place over many decades and demonstrate that fluctuating salinities, caused by excessive precipitation, are common events in the intertidal areas around the Outer Ten Thousand Islands. Some of the other islands, such as Rabbit and Jewel Key, house small fresh or brackish water pools within their interiors, the result of rain water draining from the porous sediments that comprise the land surface (Figure 0.9). All of these environments, along with the diverse geology and oceanography of the islands systems, make the Ten Thousand Islands one of Florida's greatest natural wonders and the "Crown Jewel" of Everglades National Park; where the Everglades meets the sea.

Endemism in the Ten Thousand Islands

Because of the wide-ranging swings in salinities, the fauna of the Reticulated Coastal Swamp area is impoverished and contains far fewer species than are found in the Florida Keys to the south and along the western coast of Florida to the north. Although having fewer types of marine organisms than are seen elsewhere in southern Florida, the Ten Thousand Islands area still houses a fairly rich fauna of marine and estuarine organisms, containing some of the most beautiful and interesting marine animals found anywhere in the state (many shown here in the following chapters). These hardy organisms have evolved special physiological and anatomical adaptations that allow them to inhabit areas of wildly fluctuating salinities.

Several Ten Thousand Islands mollusks have now been shown to represent species or subspecies that were unique to the island archipelagoes (Petuch and Sargent, 2011;

Petuch and Myers, 2014; Petuch and Berschauer, 2019). Being adapted to oceanographically-chaotic ecosystems, these hardy animals dominate areas where closely-related marine species cannot survive for any length of time, and they give the molluscan fauna of the Ten Thousand Islands a distinctive composition. Five of these endemic shells include:

Gastropoda

Gemophos tinctus pacei (Pace's Dwarf Whelk), Family Pisaniidae (Figures 3.9 and 3.16)

Vermicularia knorri owensi (Owen's Worm Shell), Family Turritellidae (Figure 3.9)

Naticarius verae (Vera's Moon Snail), Family Naticidae (Figure 4.22)

Cinctura tortugana foxi (Fox's Tulip Shell), Family Fasciolariidae (Figure 6.7)

Bivalvia

Mercenaria browni (Brown's Venus Clam), Family Veneridae (Figure 4.25)

Another possible new Ten Thousand Islands subspecies includes a smooth variant of the Auger Shell, *Neoterebra dislocata* (Say, 1822) (Figure 4.23). Future research may prove that this distinctive Ten Thousand Islands variant warrants its own subspecific name.

Probably the most notable Ten Thousand Islands endemic animal is the Ten Thousand Islands Raccoon, *Procyon lotor marinus* Nelson, 1930, a small, pale tan-colored subspecies of the mainland Florida Common Raccoon, *Procyon lotor* (Linnaeus, 1758). This unique Ten Thousand Islands endemic animal lives in mangrove forests, licks dew from mangrove leaves as a source of fresh water, eats oysters and crabs, and swims between the mangrove islands. These amazingly successful mammals apparently evolved from Pleistocene forest raccoons that were left behind, and stranded on the islands when sea level rose and flooded the area. Like its mollusk and crustacean prey, the Ten Thousand Islands Raccoon is uniquely suited to live and flourish in the dense mangrove jungles and isolated outer islands.

Biogeography of the Ten Thousand Islands

The entire Ten Thousand Islands area, from Cape Romano, Collier County south to Cape Sable, Monroe County, is a faunal component of the **Carolinian Molluscan Province**, an immense biotic area that extends from Cape Hatteras, along the entire eastern coast of the United States, around the Florida Keys, and throughout the entire Gulf of Mexico (Figure 0.10; see also Petuch, 2013; Petuch and Berschauer, 2020b for details on the Carolinian Province and its characteristic mollusks). This biotic province can be defined by the combined distributions of several prominent species of mollusks, all of which are abundant throughout the Ten Thousand Islands. These include: the Horse Conch *Triplofusus giganteus* (Kiener, 1840) (Figure 4.19); the Banded Tulip Shell *Cinctura hunteria* (Perry, 1811) (Figure 3.17); the Florida Fighting Conch *Strombus alatus* (Gmelin, 1791) (Figure 4.18); the Left-Handed Whelk *Sinistrofulgur sinistrum* (Hollister, 1958) (Figure 4.22); and the Campeche Venus Clam *Mercenaria campechiensis* (Gmelin, 1791) (Figure 4.24). Their presence within the island ecosystems demonstrates that the Ten Thousand Islands molluscan fauna is classically Carolinian in composition.

The Outer Ten Thousand Islands and worm shell reefs also house a number of species that live only along western Florida, Alabama, and Mississippi, in a Carolinian subdivision named the **Suwannean Subprovince** (see Petuch, 2013; Petuch and Berschauer, 2020b for a description and definition of the subprovince). Some of these classic Suwannean endemic species include the small murex shells *Vokesimurex perrugatus* (Conrad, 1846) and *Eupleura tampaensis* (Conrad, 1846) (both Figure 3.16), the Tampa Top Shell, *Calliostoma tampaense* (Conrad, 1846) (Figures 3.14 and 3.19), and the Cone Shells *Gradiconus floridanus* (Gabb, 1869), (Figure 5.9), *Gradiconus floridanus* form *floridensis* (Sowerby I, 1870) (Figure 5.10), and *Jaspidiconus stearnsi* (Conrad, 1869) (Figure 4.22). All of these Suwannean Subprovince endemics are found only along western Florida, with the farthest south collection records being at Pavilion Key in the southern section of the Ten Thousand Islands archipelago; none of these species are found in the Florida Keys. See Chapters 3 and 5 for illustrations of these classic Suwannean Province species.

The Ten Thousand Islands area also constitutes an "evolutionary hot spot", where special localized evolution has resulted in a number of endemic species with biogeographical ranges that are confined only to the area between Cape Romano and Pavilion Key. This "hot spot" area has been referred to as the **Chokoloskean Infraprovince** (named for Chokoloskee Island; see Petuch and Berschauer, 2020b) and its faunal components reflect the special types of environments that have formed there since the late Pleistocene Epoch. The previously-mentioned endemic mollusks, *Naticarius verae*, *Gemophos tinctus pacei*, *Vermicularia knorri owensi*, *Cinctura tortugana foxi*, and *Mercenaria browni*, are all restricted to the Chokoloskean Infraprovince and represent special adaptations to the chaotic environments of the Reticulated Coastal Swamps and island archipelagoes. These special endemic Everglades National Park animals will be described and discussed in their respective chapters. Within the Chokoloskean Infraprovince, we recognize six different types of marine environments, with each supporting its own set of ecosystems. These include: **Mangrove Forests** (Chapter 1), **Oyster Banks** (Chapter 2), **Vermetoherms** (Chapter 3), **Sand and Mud Flats** (Chapter 4), **Sea Grass Beds** (Chapter 5), and **Deep Channels and Offshore Areas** (Chapter 6). The most prominent and conspicuous of the organisms that live in these six different environments, some never before illustrated in a book, are highlighted throughout the following chapters.

Pictorial Overview of Ten Thousand Islands Environments

The following photographs show some of the main types of environments and conspicuous organisms that are present in the Ten Thousand Islands.

Figure 0.1 View of Chokoloskee Bay, one of the Inner Ten Thousand Islands Back Bays. Here, large reef-like bioherms, composed of the oysters *Crassostrea virginica*, *Crassostrea rhizophorae*, and *Ostreola equestris* form extensive structures that provide the hard substrate for the colonization by Red Mangrove propagules (seedlings) and the establishment of a mangrove forest.

Figure 0.2 View of the entrance to Rabbit Key Pass, near to where it connects with Chokoloskee Bay. These deep, wide tidal channels are the primary conduits between the Inner Ten Thousand Islands and Back Bays and the Outer Ten Thousand Islands and the Gulf of Mexico.

Figure 0.3 View of the Rabbit Key Pass, showing a large Osprey (*Pandion haliaetus* (Linnaeus, 1758)) nest in a dead White Mangrove tree. Ospreys, or Fish Hawks, are common throughout the Ten Thousand Islands and build their nests in the same tree year after year, often producing large prominent structures like this one. The local fishing guides often use these nests as navigational markers along many of the winding tidal channels. Photo courtesy of Michael Bruggeman.

Figure 0.4 A young Bottle-Nosed Dolphin (the porpoise *Tursiops truncatus* Montagu, 1821) playing in the wake of the senior author's boat. Often, entire pods of dolphins will "surf the wave" of boat wakes and their excited chattering and clicking can be heard above the roar of the outboard engine. Local fishing guides always mention how individual family pods of porpoises have learned to recognize the sounds of different engines, preferring to play with the boats that have the most horse power and produce the largest waves. Some pods have even learned to recognize the sounds of individual motors and individual boats, joyfully swimming over, *en masse*, to greet their favorite humans.

Figure 0.5 View of a small bay, bordered by a long white quartz sand beach, on the northeastern side of Pavilion Key, Ten Thousand Islands, Monroe County, Florida.

Figure 0.6 View of the southwestern side of Rabbit Key, Monroe County, Florida, showing the wide variety of marine habitats that are found in the shallow sandy bays that surround the island, including Black Mangrove forests, Shoal Grass beds, and open muddy sea floors.

Figure 0.7 View of the northern side of Turtle Key, at the entrance to the Rabbit Key Pass, showing the dense island vegetation composed of Black Mangroves, Railroad Vines, and Buttonwood trees.

Figure 0.8 View of the beach area along the western side of Pavilion Key, showing the immense accumulation of large mollusks, primarily the Left-Handed Whelk, *Sinistrofulgur sinistrum* (Hollister, 1958) and the Campeche and Brown's Venus Clams, *Mercenaria campechiensis* (Gmelin, 1791) and *Mercenaria browni* Petuch and Berschauer, 2019, that covers the intertidal zone. The dead mangrove tree in the distance supports a large Osprey nest.

Figure 0.9 View of a brackish water pool within the interior of Rabbit Key. Water flow from this small lagoon cascades over a low waterfall and produces a small, shallow stream that empties into the sea along the northwestern end of the island. The pool, filled with tannin-stain water from decomposing mangrove leaves, houses a large population of the Longnose Killifish, *Fundulus similis* (Baird and Girard, 1853).

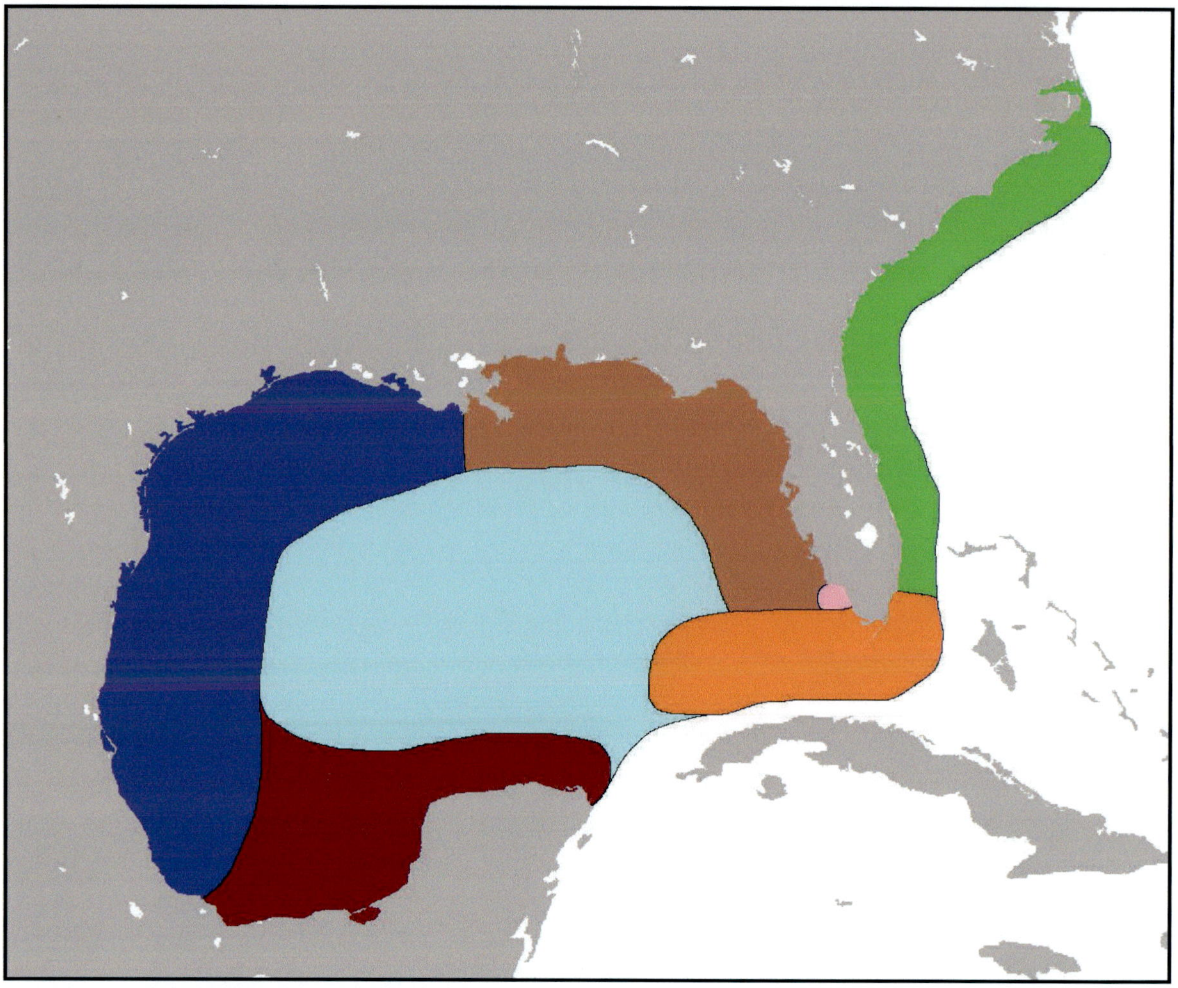

Figure 0.10 Map of the distribution of the **Carolinian Molluscan Province** in the Gulf of Mexico, showing its subprovinces: Orange= **Floridian Subprovince**; Brown= **Suwannean Subprovince**; Blue= **Texan Subprovince**; and Burgundy= **Yucatanean Subprovince**. The eastern coast of Florida is included within the **Georgian Subprovince** (Green), which extends to Cape Hatteras, North Carolina. The **Chokoloskean Infraprovince** (Pink), which encompasses the entire Ten Thousand Islands archipelago, represents a highly localized evolutionary "hot spot" where endemic species have adapted to the special ecological conditions of the area. Adapted from Petuch and Berschauer, 2020, figure 2.2. The pale blue color demarcates the central abyssal plain of the Sigsbee Deep, which contains its own endemic bathyal and abyssal fauna.

CHAPTER 1. The Ten Thousand Islands Mangrove Forests

The vast forests of intertwined tropical trees comprise the largest and most prominent type of environment seen on the archipelagoes of the Ten Thousand Islands and in the back bays. This transitional world, between land and sea, is submerged by shallow water at high tide and exposed to aerial conditions at low tide. The organisms that live in this intertidal world have evolved many amazing adaptations, allowing them to thrive in areas that would be lethal to other terrestrial or marine life. The greatest adaptions to this harsh, constantly-changing world are seen in a group of trees which have evolved the ability to live in pure salt water or in salt-saturated soils. Four separate and unrelated species, referred to collectively as **Mangrove Trees**, belonging to different tree families, have created the vast forest systems seen throughout the island chains. Four types of mangroves separate themselves by soil and salinity preferences, producing distinct vegetation zones that are dominated by one single species. Each vegetation zone supports its own distinctive ecosystem and array of marine and semi-terrestrial organisms.

Of the Mangrove Trees, the pioneer species is the **Red Mangrove** (*Rhizophora mangle*; Figure 1.1), which colonizes exposed, unvegetated sand flats, oyster banks, and worm snail reefs. The colonization process of such open and pristine areas occurs after the Red Mangrove's floating cigar-shaped seedlings (**propagules**) are washed ashore and implant themselves in mud, soft sand or oyster shell debris. Once embedded in the substrate, the Red Mangrove propagule quickly sends down deep roots and anchors the plant to the sediments. As the tree grows, it produces prominent supporting "**prop roots**" which spread out over the newly-colonized area and stabilize the sediments, preventing erosion during strong storms. The prop roots also act as debris traps, allowing shed mangrove leaves and other organic detritus to accumulate and produce a distinctive marginal marine soil called **Mangrove Peat**. The prop roots also serve as the attachment areas for the Mangrove Oyster (*Crassostrea rhizophorae*), which often form large intertwined colonies along the tide line. Dead oyster valves are also a major contributor to soil and land build-up in Red Mangrove forests, helping to elevate the entire forest system into a topographically high area that rises above the high tide level.

Once the pioneer Red Mangroves have established a large vegetated island with thick layers of mangrove peat, the central areas farthest inland from the surrounding open sea are then primed for the establishment of forests of the **Black Mangrove** (*Avicennia germinans*). The seeds of this type of mangrove tree resemble large Lima Beans and float, often in great numbers, just below the water surface and can be transported by wind and currents for great distances. At high tide, when the low mangrove islands are flooded, the Black Mangrove seeds wash into the interior of the islands and then colonize the thick peat areas. Once established, the Black Mangroves grow very quickly and overshadow the original Red Mangrove pioneers and replace them with large mats of specialized air-breathing roots. Since the mangrove peat soil is filled with sulfides and is characterized as having anoxic conditions (oxygen-poor), the Black Mangrove needed to evolve a mechanism to overcome this lack of soil gases. To accommodate an exchange of gases from the atmosphere, the Black Mangrove root mats send up a myriad of small vertically-oriented root extensions, referred to as **pneumatophores** (Figure 1.2). These air-breathing roots ensure that the tree will receive the right percentages of gases such as oxygen, nitrogen, and carbon dioxide and are crucial for the growth of a healthy Black Mangrove forest. The pneumatophores also provide habitats for a wide variety of marine organisms, including attached mussels, crabs, sponges, and a wide variety of marine algae.

The Red and Black Mangrove forests, which most often occur in close proximity on the same island and form interlocking tree canopies, also provide the habitat of many important marine and marine-affiliated vertebrates. In the Ten Thousand Islands, the most impressive marine vertebrate is the American Crocodile (*Crocodylus acutus*; Figure 1.3) which often grows to lengths of 5 m (16.4 feet) and prefers the quiet salt water creeks and channels between the mangrove islands. Hiding among the Red Mangrove prop roots, the crocodiles use stealth predation to capture large fish, swimming birds, and small swimming mammals. One of the larger prey items of the American Crocodile is the aquatic Ten Thousand Islands Raccoon (*Procyon lotor marinus*), the distinctive light-colored endemic subspecies of the larger mainland Raccoon. When paddling in open water between islands, this small raccoon is vulnerable to attacks by crocodiles, and this predation serves to control the size of the raccoon population. The Red and Black Mangrove forests also serve as the preferred habitat and rookeries for a large number of spectacular wading shore birds. Some of the more prominent of these are the Great Egret (*Ardea alba*; Figure 1.4) and the Roseate Spoonbill (*Ajaja ajaja*; Figure 1.5), which roost along the edges of the mangrove-lined tidal channels and provide a dazzling spectacle for visiting naturalists and eco-tourists. Dozens of other smaller wading and perching birds roost in these high canopies and, along with the rich fauna of sea and beach birds, clearly demonstrate that the Ten Thousand Islands contains the richest avifauna found in the entire Everglades National Park area.

Like the Red Mangrove prop roots, the Black Mangrove pneumatophores also accumulate debris and organic matter and add to the richness of the mangrove peat. This thicker inland soil layer is often saturated with rain water and is far less saline, producing the ideal conditions for the establishment of the last two types of Ten Thousand Islands mangrove trees, the **White Mangrove** (*Laguncularia racemosa* Gaertner, 1807) and the **Buttonwood Tree** (*Conocarpus erectus* Linnaeus, 1753). The White Mangrove superficially resembles the Red Mangrove but differs in having lower prop roots that are only slightly elevated off ground level and also by its much taller canopy, which frequently reaches heights of over 20 m (65.6 feet). White Mangroves are far more frequently encountered along the back bays and on the older, larger islands, where they tower over the smaller Red and Black Mangroves that are closer to the water's edge. The large, bushy Buttonwood Tree, sometimes called the Green Mangrove Tree, prefers the highest elevation areas and is only rarely flooded with sea water. This fourth type of mangrove is mostly absent from the Outer Ten Thousand Islands and is found far inland along the edges of sawgrass meadows or in the relatively dry interiors of large islands. There, it often occurs together with groves of the elongated, stick-like Dildo Cactus (*Acanthocereus tetragonus*) (note: Dildo Key in nearby Florida Bay takes its name from the slightly off-color vernacular moniker for this spiny plant!).

Mangrove-Associated Marine Organisms

The Red Mangrove trees offer two different habitats for the invertebrate organisms that coexist with them: the ecological zone of the high tree canopy, upper branches, and main tree trunk; and the ecological zone of the submerged prop roots and accompanying oyster clumps. These organisms are segregated in their distributions on the tree, with each being restricted to one of the two ecological zones. One of the more prominent and conspicuous of these arboreal invertebrates is the Mangrove Periwinkle (*Littoraria angulifera*, Figure 1.6), which often climbs high into the tree canopy, feeding on algal films that grow on the moist mangrove leaves. Lower down on the tree, primarily on the main trunk immediately above the high tide water line, a smaller relative of the Mangrove Periwinkle, the Cloudy Periwinkle, *Littoraria nebulosa* (Figure 1.18), often occurs in large aggregations (Figure 1.7). The driftwood piles along the beaches of the Outer Ten Thousand Islands are sometimes covered with immense masses of these pretty blue and gray snails and their checkered color form, *L. nebulosa* form *tessellata* (Figures 1.20 I, J). Although rare in the nearby Florida Keys, the Cloudy Periwinkle is exceptionally abundant in some Ten Thousand Islands areas. The Gulf Marsh Periwinkle, *Littoraria irrorata sayi* (Figure 1.20 H), a Gulf of Mexico subspecies of the eastern United States *L. irrorata*, is also found along with the Clouded Periwinkles on drift wood accumulations but is rare in the Ten Thousand Islands. Another of the high canopy and upper branches invertebrates is the Mangrove Tree Crab (*Aratus pisonii*, Figure 1.8). This small, spider-like crab is highly active and can be seen skittering up and down the sides of Red Mangrove trees. Within large, dense tropical hardwood forests on the larger islands, the Banded Tree Snail (*Orthalicus floridensis*; Figure 1.13) occurs higher up on the trunks and branches of salt-tolerant trees, where it is easy prey for snail-eating birds.

The intertidal mangrove root environment, collectively composed of Red, Black, and White Mangroves, has its own distinctive molluscan fauna, one which has evolved to survive extremes in ecological conditions. Being exposed to aerial conditions during low tide, these marine organisms have developed the physiological mechanisms to survive extremes in air temperature, during cold winters and hot summers, and extremes in salinity during heavy summer rains. Nestled between the algae-covered pneumatophore roots of Black Mangroves, the Granulated Mussel (*Geukensia granossisima*; Figure 1.9) often forms large, prominent aggregations and is the most frequently-encountered mollusk in the mangrove peat beds. Occurring high above the water line at low tide, these beautiful sculptured mussels are safe from most marine predators but are easy prey for the Ten Thousand Islands Raccoon (*Procyon lotor marinus*) and are one of the principal food resources for this unusual endemic semi-aquatic mammal. The Red Mangrove prop roots also house large aggregations of bivalves, in this case the specialized Mangrove Oyster (*Crassostrea rhizophorae*; Figure 1.10). This small bivalve, which has evolved a commensal relationship with the Red Mangrove, occurs only on the roots or submerged branches and forms dense aggregations. These masses of attached shells form a hard surface, and also a myriad of hiding places, for a variety of small gastropod mollusks. Some of these include air-breathing snails such as the Coffee Bean Shells (*Melampus bidentatus*, *Melampus coffea*, *Melampus monile*) and the West Indian Ear Shell (*Ellobium (Auriculoides) pellucens*) (all shown here in Figure 1.17). On the mud flats adjacent to the mangrove roots, swarms of the Sand Fiddler Crab (*Leptuca pugilator*; Figure 1.11) often cover the area at low tide and these provide a rich food resource for many of the shore birds.

Arboreal Gastropods of Island Tropical Hardwood Forests

Scattered among the labyrinth of low mangrove islands are several high, raised land masses, with the most prominent being Fakahatchee and Chokoloskee Islands and Russell Key. In pre-Columbian times, the Calusa People lived on these high islands and constantly enlarged them, intentionally modifying the landscape by repeatedly dumping empty oyster shells (after consuming their meat) around the insular perimeter, creating platform mounds, fish traps, and retention pools (Macmahon & Marquardt, 2004; Marquardt & Kozuch, 2016; Thompson *et al.*, 2016, Thompson *et al.*, 2020). Over the centuries, many of these peripheral refuse piles ("oyster middens") grew to immense proportions, greatly expanding islands such as Marco, Fakahatchee, and Chokoloskee (a shoreline oyster midden on Fakahatchee Island is shown here in Figure 1.14). Because these islands are elevated above the surrounding estuaries and tidal creeks (3-5 m), both the top layers of soil and the island interiors are salt-free and can support large forests of tropical hardwoods and delicate tropical under story plants. Some of these include Gumbo Limbo and Pigeon Plum trees, along with delicate cactus plants such as the Florida Dildo Cactus (*Acanthocereus tetragonus*; Figure 1.15). These tropical forests now reside only on the highest islands of the entire archipelago.

The tropical hardwood forests within the interiors of these high islands also house a distinctive color form variant of the Brown-Banded Everglades Tree Snail, *Liguus fasciatus castaneozonatus* Pilsbry, 1912, that is unique to the Ten Thousand Islands. Deriving from a Cuban ancestor, *Liguus fasciatus fasciatus* Muller, 1774, the progenitor of the Everglades *castaneozonatus* rafted across the Straits of Florida, attached to trees that were part of huge forest "islands" that had washed off the Cuban mountains during late Pleistocene super-hurricanes (see Close, 2000, Appendix A; Petuch and Myers, 2014: 215-217). These ancestral snails were accompanied by entire displaced forests of the Royal Palm (*Roystonea regia* (Kunth, 1816)), which also managed to survive the storm-driven sea voyage and to have established themselves in the Fakahatchee Strand area along the mainland side of Fakahatchee Bay. Populations of the ancestral Cuban *Liguus*

Tree Snails quickly colonized the Florida Keys, Ten Thousand Islands, and southern Everglades, and later, due to rising sea levels, became geographically and genetically isolated from each other. During Holocene time, these isolated populations of *Liguus fasciatus* diverged from each other and produced an amazing complex of at least 6 subspecies and over 60 named color forms.

The pale color form of *Liguus fasciatus castaneozonatus* that occurs on the high Ten Thousand Islands was only recently discovered to be morphologically-distinct from other *Liguus* populations in the Everglades region and the Florida Keys. This geographically-isolated variant, which is now known to be endemic to the Ten Thousand Islands, was given the form name *mitchelli* by Poland in 2008 (specimens shown here in Figures 1.16 and 1.21; see Poland, 2008). Since *Liguus* snails are extremely sensitive to salt, either in water or as salt spray, they always live deep within the tropical forests on large islands or in mainland hammocks, completely protected from the influences of the sea. Although being surrounded by vast areas of salt water lagoons, channels, and mangrove forests, populations of *Liguus fasciatus castaneozonatus* form *mitchelli* are now known to have survived on all the high islands of the archipelago, primarily Fakahatchee, Russell, and Chokoloskee (the type locality of the form *mitchelli* Poland, 2008 is on Chokoloskee Island, near the old cemetery) (Adrián González-Guillén, *et al.,* 2018). In the hardwood and cactus forests on the high islands, the Ten Thousand Islands endemic color form *mitchelli* lives together with the large, widespread Florida Tree Snail, *Orthalicus floridensis* (Pilsbry, 1899) (Figure 1.13) and can be considered to be a special biological treasure.

Pictorial Overview of Mangrove Environments and Associated Organisms

The following photographs show some of the main types of environments and conspicuous organisms that are present in the Mangrove Forests of the Ten Thousand Islands.

Figure 1.1 View of an established forest of Red Mangrove Trees (*Rhizophora mangle* Linnaeus, 1753) growing along the shoreline of the Rabbit Key Channel. Here, the characteristic prop roots form an intertwined mass that supports the trees and provides shelter for numerous resident animals.

Figure 1.2 View of a forest of Black Mangrove Trees (*Avicennia germinans* (Linnaeus, 1764)) growing along the southern shoreline of Turtle Key, showing a well-developed mat of pneumatophores ("air-breathing roots").

Figure 1.3 An American Crocodile (*Crocodylus acutus* Cuvier, 1807) floating among the prop roots of a Red Mangrove Tree in Ponce de Leon Bay, near the mouth of the Shark River in the southern Ten Thousand Islands. Photo courtesy of Morris A. Foster.

Figure 1.4 Four Great Egrets (*Ardea alba* Linnaeus, 1758) and a Double-Crested Cormorant roosting in a solitary Red Mangrove tree in the middle of Chokoloskee Bay, Collier County. Photo courtesy of Morris A. Foster.

Figure 1.5 A pair of Roseate Spoonbills (*Ajaja ajaja* (Linnaeus, 1758)) roosting in a large Red Mangrove tree along the south coast of Ponce de Leon Bay, near the mouth of the Shark River in the southern Ten Thousand Islands. Photo courtesy of Morris A. Foster.

Figure 1.6 Close-up view of two Mangrove Periwinkles, *Littoraria angulifera* (Lamarck, 1822), crawling on the trunk of a Red Mangrove tree on Turtle Key. This is the largest littorinid periwinkle found in the Ten Thousand Islands and it prefers the highest sections of the mangrove trees, well above the high tide line and often crawls directly on the mangrove leaves. Photo courtesy of Michael Bruggeman.

Figure 1.7 Close-up view of an aggregation of the Cloudy Periwinkle, *Littoraria nebulosa* (Lamarck, 1822), on the trunk of a dead mangrove tree on Rabbit Key. This is the most abundant periwinkle on the drift wood piles that line many of the Ten Thousand Islands beaches. Preferring the lower areas of the driftwood and living mangroves, the Cloudy Periwinkle is most frequently seen to form large aggregations along the edge of the high tide splash zone. Photo courtesy of Michael Bruggeman.

Figure 1.8 Close-up view of a Mangrove Tree Crab, *Aratus pisonii* (Milne-Edwards, 1837), crawling on the trunk of a large Red Mangrove tree along the shoreline of Chokoloskee Island, Chokoloskee Bay. Photo courtesy of Morris A. Foster.

Figure 1.9 Close-up view of algae-covered Black Mangrove pneumatophores, showing an aggregation of the mussel, *Guekensia granosissima* (Sowerby III, 1914), which are embedded between the individual roots and attached by strong byssal threads.

Figure 1.10 View of Red Mangrove (*Rhizophora mangle*) prop roots covered with clumps of the Mangrove Oyster, *Crassostrea rhizophorae* (Guilding, 1828).

Figure 1.11 Close-up view of a male Sand Fiddler Crab, *Leptuca pugilator* (Bosc, 1802), on an exposed sand bar near mangrove forests. Immense aggregations of Sand Fiddle Crabs typically swarm sand bars and mud flats that are exposed at low tide. Photo courtesy of Michael Bruggeman.

Figure 1.12 A flock of the Double-Crested Cormorant (*Phalacrocorax auritus* (Lesson, 1831), roosts in a dead mangrove tree within Chokoloskee Bay, Collier County. Photo courtesy of Michael Bruggeman.

Figure 1.13 Close-up view of the Florida Tree Snail, *Orthalicus floridensis* (Pilsbry, 1899), crawling on a Buttonwood Tree on Pavilion Key, Ten Thousand Islands, Monroe County, Florida. These large, conspicuous tree snails are common on large islands that support tropical hardwood forests and are also abundant throughout the southern region of the Everglades.

Figure 1.14 View of the south shore of Fakahatchee Island, Fakahatchee Bay, Ten Thousand Islands, Collier County, Florida, showing the eroded edge of the huge oyster shell midden. These stratified layers of oyster valves were deposited as refuse piles by the Calusa People and demonstrate the importance of oysters in the diet of the regional Native Americans. The tropical hardwood forests on Fakahatchee Island, which grow on top of these oyster middens, house a large population of the endemic Ten Thousand Islands Tree Snail, *Liguus fasciatus castaneozonatus* form *mitchelli* Poland, 2008.

Figure 1.15 View of the interior of Fakahatchee Island, Fakahatchee Bay, Ten Thousand Islands, Collier County, Florida, showing the dense tropical and subtropical vegetation that grows within the higher areas of the island. The extremely spiny Florida Dildo Cactus, *Acanthocereus tetragonus* (Linnaeus, 1753), can be seen to dominate much of the understory vegetation of the island's interior forests.

Figure 1.16 Close-up view of a living specimen of the Ten Thousand Islands Tree Snail, *Liguus fasciatus castaneozonatus* form *mitchelli* Poland, 2008, crawling on a low shrub growing within the tropical hardwood forest on Fakahatchee Island, Fakahatchee Bay, Ten Thousand Islands, Collier County, Florida. This pale-colored variant of the dark-banded *Liguus fasciatus castaneozonatus* Pilsbry, 1912, is found only in the Ten Thousand Islands archipelago and is most common on Fakahatchee and Chokoloskee Islands and on Russell Key. The form *mitchelli* exists as a pure colony, with no other *Liguus* color forms being present and with all individuals showing little variation in shape and color pattern.

Iconography of Mangrove-Associated Mollusks

The gastropod and bivalve mollusks illustrated here are all classic inhabitants of the Ten Thousand Islands mangrove forests. Some of the most important ecological index taxa include:

Gastropoda
Littorinidae
Littoraria angulifera (Lamarck, 1822)
Littoraria irrorata sayi (Philippi, 1846)
Littoraria nebulosa (Lamarck, 1822)
Littoraria nebulosa form *tessellata* (Philippi, 1847)
Melampidae
Melampus bidentatus (Say, 1822)
Melampus monilis (Bruguiere, 1789)
Melampus coffea. (Linnaeus, 1758)
Ellobiidae
Ellobium (Auriculoides) pellucens (Menke, 1830)

Arboreal Snails in an Isolated Island Hardwood Forest
Orthalicidae
Liguus fasciatus castaneozonatus form *mitchelli* Poland, 2008
Orthalicus floridensis (Pilsbry, 1899)

Bivalvia
Mytilidae
Geukensia granosissima (Sowerby III, 1914)
Ostreidae
Crassostrea rhizophorae (Guilding, 1828)

Figure 1.17 The Mangrove Periwinkle, *Littoraria angulifera* (Lamarck, 1822), showing its color variations.

A, B= dark-colored specimen with oblique bands, length 19.2 mm; **C, D**= light-colored specimen with oblique bands, length 21.7 mm; **E, F**= pale tan specimen, length 22.3 mm; **G**= yellow-tan specimen, length 23.3 mm; **H**= specimen with a typical color pattern, length 19.8 mm.

Figure 1.18 The Cloudy Periwinkle, *Littoraria nebulosa* (Lamarck, 1822), showing its color variations.
A, B= specimen with dark-colored spire whorls, length 17.2 mm; **C, D**= typical specimen, length 20.5 mm; **E, F**= less common pale-colored specimen, length 17.5 mm; **G**= specimen with the early whorls having a checkered color pattern, length 14.8 mm; **H**= specimen with the early whorls having a checkered color pattern, length 13.4 mm.

Figure 1.19 Bivalves of the Mangrove Root environments.
A, B= *Geukensia granossisima* (Sowerby III, 1914), length 59 mm; **C=** *Crassostrea rhizophorae* (Guilding, 1828), length 58 mm; **D=** *Crassostrea rhizophorae* (Guilding, 1828), elongated form, length 132 mm; **E=** *Ostreola equestris* (Say, 1834), length 61 mm; **F=** *Ostreola equestris* (Say, 1834) growing on a 140 mm aggregation of the worm shell *Petaloconchus nigricans* (Dall, 1884).

Figure 1.20 Gastropods of the Mangrove Root environments.
A, B= *Melampus coffea* (Linnaeus, 1758), length 17 mm; **C, D=** *Melampus monilis* (Bruguiere, 1789), length 16 mm; **E, F=** *Ellobium (Auriculoides) pellucens* Menke, 1830, length 23 mm; **G=** *Ellobium (Auriculoides) pellucens* Menke, 1830, length 25 mm; **H=** *Littoraria irrorata sayi* (Philippi, 1846), length 13 mm (the Gulf of Mexico subspecies of the eastern United States *Littoraria irrorata* (Say, 1822); not found in the Florida Keys and rare in the Ten Thousand Islands); **I, J=** *Littoraria nebulosa* form *tessellata* (Philippi, 1847), length 15 mm; **K, L=** *Melampus bidentatus* (Say, 1822), length 11 mm.

Figure 1.21 Arboreal Gastropods of the Hardwood Forest on Fakahatchee Island; Comparison Specimens from Chokoloskee Island and Russel Key.
A, B, C= *Liguus fasciatus castaneozonatus* form *mitchelli* Poland, 2008, from Fakahatchee Island, voucher specimen LACM 3783, length 51.5 mm (with red patch on the edge of the lip); **D, E, F**= *Liguus fasciatus castaneozonatus* form *mitchelli* Poland, 2008, typical specimen from Fakahatchee Island, length 48 mm (with red patch on the edge of the lip); **G**= *Liguus fasciatus castaneozonatus* form *mitchelli* Poland, 2008, holotype from Chokoloskee Island (Phil Poland photo courtesy of Adrián González-Guillén and Pete Krull); **H**= *Liguus fasciatus castaneozonatus* form *mitchelli* Poland, 2008, from Fakahatchee Island (photo courtesy of Adrián González-Guillén and Pete Krull); **I**= *Liguus fasciatus castaneozonatus* form *mitchelli* Poland, 2008, from Russell Key (photo courtesy of Adrián González-Guillén and Pete Krull).

Figure 1.22 Ten Thousand Islands Raccoon. Close-up of three individuals of the Ten Thousand Islands Raccoon, *Procyon lotor marinus* Nelson, 1930, climbing on Red Mangroves. This semi-aquatic marine subspecies is unique to the Ten Thousand Islands where it swims between mangrove islands in search of mussel, oyster and crustacean prey. They build their dens among the mangrove roots and lick dew and raindrops from mangrove leaves as a source of drinking water. Photo courtesy of Captain Craig Daniels.

CHAPTER 2. The Ten Thousand Islands Oyster Banks

After the mangrove forests, the reef-like bioherms produced by oysters ("Oyster Banks") are the most prominent biologically-produced structures in the Ten Thousand Islands. Often covering hundreds of acres within the lagoons, Back Bays, and tidal channels, these banks are composed primarily of two species of oysters: the Virginia Oyster (*Crassostrea virginica*; Figures 2.1 and 2.2); and the Crested Oyster (*Ostreola equestris*; Figure 2.3). Occurring in large clumps that project above the water level at low tide, the Virginia Oyster is most prominent in the higher-salinity lagoons and tidal channels. Preferring brackish water, lower-salinity muddy areas near mangrove forests, the Crested Oyster only grows together with the Virginia Oyster in the Back Bays and in areas where there are wildly-fluctuating salinities due to river effluent and strong tidal currents.

Within the Back Bay lagoons, such as in Chokoloskee Bay (shown here in Figures 2.1 and 2.2), the oyster bars form massive biohermal structures that often extend for tens of meters on a side. These biogenic landforms form the perfect substrate for the colonization of Red Mangrove trees. Here, floating Red Mangrove propagule seedlings become lodged in the oyster shells and debris of the banks and quickly take root. After a few years, these small trees grow into entire mangrove forests that completely cover their oyster bank base. Essentially every mangrove island within the Inner Ten Thousand Islands has formed in this manner, later growing together to form larger, more extensive mature mangrove forests. The original Native American inhabitants of the Ten Thousand Islands, especially the Calusa People, utilized the rich oyster food resources found on the Back Bay banks and left behind immense refuse heaps of empty shells ("middens"). Over time, these shell middens became so large and topographically-high that small cities were actually built on top of them. Chokoloskee Island represents one of these Calusa shell midden islands within Chokoloskee Bay, as does Fakahatchee Island in Fakahatchee Bay. The oyster banks also provide safe roosting areas for many types of shore birds and waterfowl. One of the more prominent and conspicuous birds which utilize the oyster banks as rookeries is the large American White Pelican (*Pelecanus erythrorhynchos*; Figure 2.4), which only visits the Ten Thousand Islands during the winter months.

The principal predator on the oyster banks and adjacent shoreline areas is a conspicuous spiny gastropod referred to as the "King's Crown Conch" (*Melongena (Rexmela) corona* (Gmelin, 1791); Figures 2.5, 2.6 and 2.10 show typical individuals). This mollusk ranges throughout the entire archipelago and exhibits several distinct shell varieties, depending on the salinity and substrate type of the area. Individuals that live on soft, muddy substrates near oyster clumps are generally smaller than normal King's Crown Conchs and develop extremely long and delicate spines, which often occur in multiple rows. This spinose form was given the name *Melongena (Rexmela) corona* form *perspinosa* by Pilsbry and Vanatta in 1934 and this is the typical variant found along the western side of Chokoloskee Island (Figures 2.7, 2.8, and 2.11). Some specimens of the *perspinosa* form develop four or five rows of spines, making them the most ornate member of their entire family (the Melongenidae, found in tropical areas worldwide). These extremely spiny Crown Conchs, which are found primarily around Chokoloskee Island, are the most distinctive gastropod found in the quiet, sheltered Back Bay environments.

King's Crown Conchs that live on quartz sand and oyster shell rubble are often dwarfed, and have poorly-developed spines, usually composed of a single row around the shoulder and spire. This dwarf form was given the name *Melongena (Rexmela) corona* form *belknapi* by Petit de la Saussaye in 1852 (who originally considered it to be a separate species; Figures 2.9, 2.13). This distinctive dwarf form, and others with three rows of knobs (form *trinodosa*; Figure 2.12) and webbed varices (Figure 2.12), all feed on juvenile oysters by forcing the shell valves apart with their powerful foot and then inserting their tooth-filled proboscis into the open oyster shell and rasping out chunks of living tissue. The King's Crown Conchs also feed upon three different species of Slipper Shells (family Crepidulidae), which attach themselves to the interiors of dead oyster valves. There, these flattened gastropods (*Crepidula fornicata*, *Ianacus atrasolea*, and *Ianacus plana*; Figure 2.14) live a sessile lifestyle and filter the sea water for particles of food, much like bivalved mollusks. The Slipper Shells are easily detached from their oyster valve residences and frequently fall victim to melongenid predation. Besides preying upon living oysters and slipper shells, *Melongena (Rexmela) corona* is also a scavenger, congregating in large numbers to feed on dead and rotting sea life (as shown in Figure 4.12 in Chapter 4 of this book).

Pictorial Overview of Oyster Bank Environments and Associated Organisms

The following photographs show some of the main types of environments and conspicuous organisms that are present on the oyster banks of the Ten Thousand Islands.

Figure 2.1 View of an oyster bank in Chokoloskee Bay, Collier County, showing aggregations of the Crested Oyster, *Ostreola equestris* (Say, 1834) and the Virginia Oyster, *Crassostrea virginica* (Gmelin, 1791). The Virginia Oysters prefer slightly deeper water and occur in large clusters on the open muddy sea floor areas.

Figure 2.2 Close-up view of a Red Mangrove forest that has grown on a large oyster bank in Chokoloskee Bay, creating the beginnings of a stable mangrove island. The oysters are pioneer organisms, building up topographic highs and stable hard-surface structures in shallow water areas. These, in turn, provide Red Mangrove propagules with a substrate on which to grow and act as the foundation for mangrove island development.

Figure 2.3 Close-up view of aggregations of the Crested Oyster, *Ostreola equestris* (Say, 1834) and the Recurved Mussel, *Ischadium recurvum* (Rafinesque, 1820) in Chokoloskee Bay. These two sessile (attached) bivalves are the principal prey of the Florida Crown Conch, *Melongena (Rexmela) corona* (Gmelin, 1791), and its many named varieties and forms.

Figure 2.4 A flotilla of the American White Pelican, *Pelecanus erythrorhynchos* Gmelin, 1789, resting on an oyster bank during low tide in Chokoloskee Bay. These large, impressive birds are seasonal residents in the Ten Thousand Islands, spending the winter months within the lagoons and back bays and migrating in early spring to their breeding grounds in the northwestern United States and south-central Canada. Photo courtesy of Michael Bruggeman.

Figure 2.5 Close-up view of the Florida Crown Conch, *Melongena (Rexmela) corona* (Gmelin, 1791), exposed at low tide in Chokoloskee Bay. The shell is covered with mud and algal films and the long, white siphon of the living animal can be seen projecting from the short siphonal canal at the anterior end. These active, carnivorous snails sense their prey by chemical signals and most often feed on thin-shelled juvenile oysters and small attached bivalves. Crown Conchs also feed on carrion and will form large aggregations that gather around dead fish and Horseshoe Crabs. Photo courtesy of Michael Bruggeman.

Figure 2.6 Close-up view of a female Florida Crown Conch, *Melongena (Rexmela) corona* (Gmelin, 1791), with her freshly-laid egg capsules. Each capsule contains around 25 larval snails, and these hatch directly into miniature adults which crawl off immediately in search of prey. Photo courtesy of Michael Bruggeman.

Figure 2.7 Close-up view of the extremely spiny variant of the Florida Crown Conch, *Melongena (Rexmela) corona* form *perspinosa* Pilsbry and Vanatta, 1934, crawling on an aggregation of the Crested Oyster. This rare and beautiful variant of the Florida Crown Conch is only found in a few quiet, protected lagoons along the Ten Thousand Islands, where it occurs on mud flats near mangrove forests.

Figure 2.8 View of a clump of Virginia and Crested Oysters, showing an aggregation of six *Melongena (Rexmela) corona* form *perspinosa* Pilsbry and Vanatta, 1934, in Chokoloskee Bay. The carnivorous Crown Conchs are resting during low tide but will begin actively crawling about and feeding when deeper water conditions return at high tide.

Figure 2.9 View of a clump of Mangrove Oysters growing on a Red Mangrove prop root, showing an aggregation of four *Melongena (Rexmela) corona* (Gmelin, 1791) dwarf form *belknapi* Petit de la Saussaye, 1852, on Turtle Key. This dwarf variety, with reduced spine development, is typically found in mangrove forests along the Outer Ten Thousand Islands.

Iconography of Oyster Bank-Associated Mollusks

The gastropod and bivalve mollusks illustrated here are all classic inhabitants of the Ten Thousand Islands oyster banks. Some of the more important ecological index taxa include:

Gastropoda
Crepidulidae
Crepidula fornicata (Linnaeus, 1758)
Ianacus atrasolea (Collins, 2002)
Ianacus plana (Say, 1822)
Melongenidae
Melongena (Rexmela) corona (Gmelin, 1791) (a polytypic species with several named forms)
Melongena (Rexmela) corona form *belknapi* Petit de la Saussaye, 1852
Melongena (Rexmela) corona form *perspinosa* Pilsbry and Vanatta, 1934
Melongena (Rexmela) corona form *trinodosa* Emery and Lermond, 1936

Bivalvia
Mytilidae
Ischadium recurvum (Rafinesque, 1820)
Anomiidae
Anomia simplex (d'Orbigny, 1853)
Ostreidae
Crassostrea virginica (Gmelin, 1791)
Ostreola equestris (Say, 1834)

Figure 2.10 The Florida Crown Conch, *Melongena (Rexmela) corona* (Gmelin, 1791), Typical Shell Forms.

A, B= specimen length 78 mm; **C, D**= specimen length 73 mm; **E, F**= specimen length 49 mm; **G, H**= specimen length 46 mm. This shell variety represents the classic Crown Conch shape, size, and number of rows of spines, and is abundant along all the Outer Ten Thousand Islands and the west coast of Florida as far north as Cedar Key.

Figure 2.11 The Florida Crown Conch, *Melongena (Rexmela) corona* (Gmelin, 1791), Extreme Spiny Form *perspinosa* Pilsbry and Vanatta, 1934.

A, B= typical spiny variety, length 65 mm; **C, D**= pale-colored variety, length 56 mm; **E, F**= typical spiny variety, length 60 mm; **G, H**= typical spiny variety, length 61 mm. This hyper-spinose variant is found only on oyster clumps in areas of soft, flocculent mud in the Back Bay lagoons. The *perspinosa* populations on the bay side of Chokoloskee Island have the best-developed spines and are one of the most spectacular mollusks found in the Ten Thousand Islands.

Figure 2.12 Variants of the Florida Crown Conch, *Melongena (Rexmela) corona* (Gmelin, 1791).

A, B= form *trinodosa* Emery and Lermond, 1936, length 62 mm; **C, D**= form *trinodosa* Emery and Lermond, 1936, length 53 mm; **E, F**= form with webbed varices, length 51 mm; **G, H**= form with webbed varices, length 50 mm. These two variants occur in the same muddy areas as does the *perspinosa* form, but are not as frequently encountered.

Figure 2.13 The Florida Crown Conch, *Melongena (Rexmela) corona* (Gmelin, 1791), dwarf form *belknapi* Petit de la Saussaye, 1852.
A, B= large specimen, length 53 mm; **C, D**= typical dwarf specimen, length 51 mm; **E, F**= typical dwarf specimen, length 43 mm. This variant is found primarily on the Outer Ten Thousand Islands, where it is associated with oyster clumps near Red Mangrove roots.

Figure 2.14 Sessile Gastropods and Bivalves of the Oyster Banks.
A, B= *Crepidula fornicata* (Linnaeus, 1758), length 39 mm; **C, D=** *Ianacus atrasolea* (Collin, 2002), length 28 mm; **E, F=** *Ianacus plana* (Say, 1822), length 21 mm; **G=** *Anomia simplex* d'Orbigny, 1853, length 40 mm; **H=** *Ischadium recurvum* (Rafinesque, 1820), length 39 mm. All of these species attach themselves to the inside of dead oyster valves, often in large aggregations, and are principal prey items of the carnivorous King's Crown Conch.

View of a vermetoherm on Turtle Key.

CHAPTER 3. The Vermetoherms of the Ten Thousand Islands

The intertidal areas of the Outer Ten Thousand Islands support gigantic biohermal structures, some covering acres of shoreline, and nothing comparable to these features has ever been found anywhere else along Florida or the Gulf of Mexico. Composed entirely of a monoculture of the Dark Worm Shell, *Petaloconchus nigricans* (Dall, 1884) (Figure 3.15), these snail shell aggregations grow together in fused masses and actually create true reef-like bioherms, referred to here as **Vermetoherms**, and are the only-known large-scale gastropod reef systems found anywhere on Earth (Shier, 1969; Petuch and Myers, 2014). Recent research has indicated that *Petaloconchus nigricans* may actually represent a small-tubed, highly-intertwined variant, or ecomorph, of the common and widespread western Atlantic *Petaloconchus varians* (d'Orbigny, 1839) and not a full species all to itself. Detailed anatomical and biochemical studies are definitely needed to resolve this taxonomic problem, but in this book we will retain the traditional name of *Petaloconchus nigricans*, as the intertwined shell masses attributed to that taxon are distinctive and may represent a local subspecies that is endemic to western and southwestern Florida. These massive reef systems can be seen along the seaward sides of larger keys such as Pavilion, Rabbit, Turtle (Figure 3.1), Demijohn, and Jewell (Figure 3.2) and are best-developed on the islands adjacent to the Fakahatchee, Chokoloskee, and Rabbit Key Passes.

All vermetid gastropods are suspension feeders, with individuals exuding a mucus net from their foot and capturing particles of food on these mucus threads. Most of their food is composed of plankton and suspended organic matter, ensnared from the water column when the reefs are submerged at high tide. When the mucus nets are covered with particles, the worm snail pulls the net in and eats it, ingesting both the mucus (which is recycled) and the ensnared plankton. The oceanographic and hydrologic conditions of the Ten Thousand Islands have produced the "perfect storm" for the formation of vermetoherms complexes, creating ecological conditions along the outer islands that are uniquely suited for massive reef development. The Outer Ten Thousand Islands are far enough away from the Everglades shoreline to ensure consistently high salinities, but yet are bathed by the nutrient-rich fresh water effluent emanating from the adjacent Everglades river mouths. The mixed water conditions, and unique water chemistry, support vast plankton resources and these produce enough food to allow the worm shell reefs to grow to exceptional sizes.

Vermetoherm Platforms and Their Complex Structure

Like their cnidarian coral reef analogues, the vermetoherms also have deterministic growth with a distinct zonated structure (see Petuch and Berschauer, 2021 for a detailed description of Florida coral reef zonation). As seen here in Figures 3.1 and 3.2, the larger and best-developed vermetoherms form wide platforms that grow seaward into the areas of higher-energy wave action. This preferred growth produces broad zones with distinct morphologies, consisting of a **Fore-Reef Area**, a main **Reef Platform**, and a **Back-Reef Lagoon**. These three ecological zones are present on all the vermetoherms found throughout the Outer Ten Thousand Islands. The degree of development of the three reef zones, on any given "reef", generally correlates with the age of the entire structure; the reefs with the widest and best-developed zones often being the oldest and the reefs with the least-developed zones being the youngest. Living worm shells are mostly confined to the outer edge of the Fore-Reef, while much of the Reef Platform and Back-Reef Lagoon aggregations are dead and may represent subfossil masses that date from early Holocene time, approximately 11,000 years ago. Many of the dead sections on the Reef Platform and Back-Reef Lagoon were previously buried by Black Mangrove forests; these had grown on top of the original early Holocene-aged vermetid aggregations but are now being uncovered and exposed by erosion due to the present rising sea levels.

Along the Fore-Reef area, adjacent to the outer edge of the main platform, the *Petaloconchus* aggregations take on a unique, flattened, sheet-like growth form (Figures 3.3 and 3.4). This new type of vermetid reef structure is here referred to as a **Stromatoverm** ("layered worm aggregation"). Growing directly into the waves along the edge of the reef, stromatoverms offer less resistance to strong wave surge, especially during storms. In the Florida Keys and Caribbean Region of today, the Elkhorn Coral, *Acropora palmata* (Lamarck, 1816), forms similar large flattened sheets and also grows directly into strong wave surge and currents at the edge of the large zonated reefs. Both the stromatoverms of the *Petaloconchus* **Fore-reef** and the flattened sheet-like growth form of *Acropora palmata* Elkhorn Corals are completely hydrodynamic, presenting thin blade-like surfaces to the breaking waves and allowing the wave surge to flow above and below the flattened vermetid aggregations and coral colonies. Because of this growth form adaptation, fragile organisms like the worm shells and corals can live and flourish in high-energy areas and surge zones. Local variations in the shape and structure of the Fore-Reef stromatoverms has also been observed, with those on some vermetoherms, like Turtle Key (Figure 3.3) forming thickened laminar structures and others, like those on Jewell Key (Figure 3.4), forming thin, structures resembling overlapping scales.

Another prominent worm shell growth form, one comprising wide concentric circles that are arranged in a "bulls-eye" pattern, occurs only on the main platform. This new type of vermetid growth structure is here referred to as a **Circoverm** ("circular worm aggregation"). Forming well away from the flattened sheets along the seaward edge, circoverms can occur as single, stand-alone features on the top of the main worm shell reef platform (Figure 3.5) or as aggregations of numerous concentric circles that fuse together in an interconnected labyrinth (Figure 3.6). In the quiet water area behind the Fore-Reef zone, these circoverm "bulls-eyes" eventually grow together to create the solid vermetid pavement that typifies the surface of the main vermetoherm platform. On some islands, like Pavilion, Rabbit, Lumber, and Turtle Keys, single isolated circoverms are most commonly encountered, and these often grow to large sizes that are several meters across. On other islands, such as Jewell, Demijohn, Jack Daniels, and Camp Lulu, the circoverms are smaller and more numerous and commonly occur as closely-packed, interconnected networks of curved arcs and semicircles.

Behind the Reef Platform Zone of Circoverms, and adjacent to the island shoreline, a lower, depressed area often develops and characteristically remains filled with sea water at low tide. This flooded area constitutes the Back-Reef Lagoon and offers a wide variety of habitats for marine organisms (Figure 3.7). The excess water that is trapped in the reef lagoon at low tide drains back into the ocean through a series of interconnected narrow channels that crisscross the main platform, with the channels typically lined with living worm shells. The adjacent beach areas of islands with well-developed vermetoherm complexes are characteristically composed of quartz sand mixed with large amounts of broken and crushed pieces of *Petaloconchus* shells. Strong wave action during storms and hurricanes breaks off large areas of the Fore-Reef and transports large fragments up onto the island beaches, where they further disintegrate into a unique type of worm shell sand (Figure 3.8). These types of worm shell sand beaches are best developed on the central and southern Outer Ten Thousand Islands, such as Jewell, Comer, Indian, Turtle, Rabbit, and Pavilion Keys.

On the entire vermetoherm complex, the reef lagoon area houses the richest fauna of invertebrates and has the highest biodiversity. The worm shell sand shorelines and the adjacent Back-Reef Lagoon, together, provide the habitats for several large and conspicuous carnivorous gastropods. Living under large slabs of worm shells growing within the lagoon, the endemic Pace's Dwarf Whelk (*Gemophos tinctus pacei* Petuch and Sargent, 2011) is a common faunal component and is one of the main predators on the living vermetid gastropods and also on encrusting invertebrates such as cheilostome ectoprocts ("bryozoans") and hydroids (Figure 3.9). This subspecies of the wide-ranging western Atlantic pisaniid *Gemophos tinctus* (Conrad, 1846) differs from typical specimens in having a coarser beaded shell sculpture and in having a much paler shell color, typically infused with large patches of pale blue. The small Keyhole Limpets, *Diodora listeri* (d'Orbigny, 1842) and *Lucapinella suffusa* (Reeve, 1850) (both shown here in Figure 3.16), occur along with *Gemophos tinctus pacei* on the undersides of vermetid slabs, where they graze on algal films. Other large and conspicuous gastropods that are found in the Back-Reef Lagoon include the Banded Tulip Shell, *Cinctura hunteria* (Perry, 1811) (Figures 3.10 and 3.17) and the True Tulip Shell, *Fasciolaria tulipa* (Linnaeus, 1758) (Figures 3.11 and 3.18), both belonging to the family Fasciolariidae but with *C. hunteria* being restricted to the Carolinian Molluscan Province and with *F. tulipa* having a much wider range that includes the both the Carolinian and Caribbean Molluscan Provinces. The Ten Thousand Islands populations of *F. tulipa* are unusually colorful, exhibiting tones of bright red, brown, orange, blue, and purple, with solid red individuals being the predominant color form found on the vermetoherms off Turtle and Rabbit Keys (Figure 3.11). Both species feed on other mollusks and are normally direct ecological competitors with each other. As seen in Figures 3.10 and 3.11, the Back-Reef lagoon area is filled with immense aggregations of the small, black batillariid gastropod, *Lampanella minima* (Gmelin, 1791) and these are the principal prey of the two species of Tulip Shells. This nearly-infinite food resource allows the two ecologically-equal predators to co-exist without any direct competition; a classic example of how the Competitive Exclusion Principle (of Gause, 1934) can be over-ridden by an abundant food resource.

In many reef lagoons, sponges of the genera *Clathria*, *Ircinia*, and *Haliclona* can be found in tide pools and growing on oysters, and these add unexpected flashes of color to a normally-drab environment. In many cases, the brilliant tones of red, orange, yellow, blue, and purple from these sponges add a visual vibrancy to the brown and green colors of the worm shell clumps and algal growths (Figures 3.13 and 3.14). On some of the larger worm shell reef systems, these sponges form dense biohermal structures ("sponge beds") that fill large areas of the lagoon and support their own separate ecosystem (Figure 3.12), providing food resources for the Tampa Top Shell, *Calliostoma tampaense* (Conrad, 1846) (Figures 3.14 and 3.19), one of the most conspicuous inhabitants of the sponge bioherms. Larger and more extensive sponge bioherms also occur in deeper water along the edge of the fore-reef, and these sometimes extend seaward for kilometers, growing on the adjacent offshore sand banks. These deeper water sponge bioherms also support large populations of the Deer Cowrie, *Macrocypraea (Lorenzicypraea) cervus* (Linnaeus, 1771) (Figure 6.5 in Chapter 6 of this book), which often occur as dwarf individuals. The large sponge bioherms also provide a substrate for the uncoiled worm turritellid gastropods, *Vermicularia fargoi owensi* Petuch and Myers, 2014 (Figure 3.19), with the individual worm turritellids spending their entire lives embedded within the body of a single sponge. This Ten Thousand Islands endemic subspecies differs from typical *Vermicularia fargoi* Olsson, 1951, from the Gulf of Mexico and northern Cuba, in having smaller and smoother early whorls (the "turritellid growth stage") that lack the typical three strong spiral cords, and in having a more slender body whorl that is much more uncoiled and tube-like.

Pictorial Overview of Vermetoherm Environments and Associated Organisms

The following photographs show some of the main growth forms of the Ten Thousand Islands worm shell reefs, and some of the more conspicuous associated organisms. The two prominent growth forms found on the vermetoherms, the laminar **Stromatoverms** and the concentric **Circoverms**, are shown here for the first time.

Figure 3.1 View, at low tide, of the western side of Turtle Key, Ten Thousand Islands, showing an immense reef-like structure that is composed entirely of the vermetid worm gastropod, *Petaloconchus nigricans* (Dall, 1884). Individual snails cement themselves to the sea floor as larvae and then grow long, tube-like shells which become intertwined with those of other individuals, forming interconnected masses. Over time, these clumps of worm shells increase in size and form reef-like hard structures which create habitats for a myriad of other marine organisms. The *Petaloconchus nigricans* vermetoherms are unique to the Ten Thousand Islands and are best developed in the southern area of the archipelago.

Figure 3.2 View of Jewell Key, Ten Thousand Islands, at low tide, showing the extensive vermetoherm composed entirely of the vermetid worm gastropod, *Petaloconchus nigricans* (Dall, 1884), that occurs along the entire northwestern side of the island.

Figure 3.3 Close-up view of the seaward edge of the vermetoherm off Turtle Key, showing the layered structure of the **Stromatoverms**. This distinctive growth form is found only along the seaward edge of the reef systems, where the snail aggregations respond to wave action by developing flattened, blade-like layers that dissipate the wave energy without damaging their structure. This Zone of Stromatoverms constitutes the **Fore-Reef Area** of the vermetoherm.

Figure 3.4 Close-up view of the main vermetoherm off Jewell Key, Ten Thousand Islands, showing numerous well-developed flattened, blade-like sheets (**Stromatoverms**) growing along the seaward edge of the reef complex. This Zone of Stromatoverms constitutes the **Fore-Reef Area** of the Jewell Key vermetoherm complex.

Figure 3.5 Close-up view of a *Petaloconchus nigricans* aggregation on Rabbit Key, Ten Thousand Islands, which has formed a large, solitary **Circoverm**. This type of concentric growth form is frequently seen on the main platform of older and better-vermetoherms.

A **Circoverm** on Camp Lulu Key.

Figure 3.6 Close-up view of several **Circoverm** aggregations that are growing on the main vermetoherm platform on Jewell Key, Ten Thousand Islands. These concentric growth forms often combine to form circoverm clusters in areas with quiet water and less wavy conditions, creating a zone of interconnected curving arcs.

Figure 3.7 View of the **Back-Reef Lagoon** system behind the main vermetoherm complex on Rabbit Key, Ten Thousand Islands. These shallow lagoons remain filled with sea water at low tide and offer a variety of habitats that support a large fauna of marine mollusks. The excess water that is trapped in the reef lagoon drains back into the ocean through a series of interconnected narrow channels that crisscross the main platform.

Figure 3.8 View of the beach on Jewell Key, Ten Thousand Islands, Collier County, showing numerous fragments of *Petaloconchus nigricans* aggregations that have eroded off the main vermetoherm platform around the edges of the island. Many of the Outer Ten Thousand Islands beaches are made up almost entirely of crushed and pulverized worm shells, the result of erosion of the adjacent reefs during heavy storms and strong wave action.

Figure 3.9 Close-up view of the underside of a large slab of *Petaloconchus nigricans* Worm Shells, from the back-reef lagoon on Turtle Key, Ten Thousand Islands. Here, three specimens of Pace's Dwarf Whelk, *Gemophos tinctus pacei* Petuch and Sargent, 2011, find refuge from the sun at low tide, along with the small green crab *Pitho aculeata* (Gibbes, 1850) (to the left of the three dwarf whelks) and the small red Snapping Shrimp, *Synalpheus* species, (lower right). Pace's Dwarf Whelk is one of the principal predators on the sessile vermetid worm shells.

Figure 3.10 Close-up view of two Banded Tulip Shells, *Cinctura hunteria* (Perry, 1811) in a tide pool on the vermetoherm off Turtle Key, Ten Thousand Islands. The Tulip shells share the tide pool with numerous individuals of the small black Least Horn Shell, *Lampanella minima* (Gmelin, 1791), which feed on algal and bacterial films.

Figure 3.11 Close-up view of a red color form of the True Tulip Shell, *Fasciolaria tulipa* (Linnaeus, 1758), in a tide pool on the vermetoherm off Turtle Key, Ten Thousand Islands. Numerous specimens of the small black Least Horn Shell, *Lampanella minima* (Gmelin, 1791), also share the pool with the large carnivorous Tulip Shell.

Figure 3.12 View of a sponge bank ("sponge reef") growing along the seaward edge of the vermetoherm off Turtle Key, Ten Thousand Islands. These massive aggregations of sponges grow on the muddy sand areas adjacent to the main reefs and are composed of several types of sponges, including species in the genera *Ircinia*, *Haliclona*, and *Clathria*.

Figure 3.13 Close-up view of the Red Beard Sponge, *Clathria prolifera* (Ellis and Solander, 1786), growing in a tide pool on the vermetoherm off Turtle Key, Ten Thousand Islands. Red Beard Sponges are common on many of the vermetoherms and sponge banks along the southern Ten Thousand Islands.

Figure 3.14 Close-up view of an oyster clump at low tide on Turtle Key, Ten Thousand Islands, showing the resident ecosystem composed of barnacles (*Balanus* species), bright yellow *Haliclona* sponges, the black Least Horn Shell, *Lampanella minima* (Gmelin, 1791), and a rose-colored Tampa Top Shell, *Calliostoma tampaense* (Conrad, 1846). This invertebrate assemblage is very typical of those found in the back-reef lagoon areas of the vermetoherm complexes.

Iconography of Vermetoherm-Associated Mollusks

The gastropod and bivalve mollusks illustrated here are all classic inhabitants of the Ten Thousand Islands vermetoherms and the associated sponge bioherms.

Ecological index species for the Vermetoherms include:

Gastropoda
Vermetidae
Petaloconchus nigricans (Dall, 1884)
Pisaniidae
Gemophos tinctus pacei Petuch and Sargent, 2011
Muricidae
Eupleura tampaensis (Conrad, 1846)
Vokesinotus perrugatus (Conrad, 1846)

Ecological Index Species for the Reef-Associated Sponge Bioherms include:

Gastropoda
Calliostomatidae
Calliostoma tampaense (Conrad, 1846)
Turritellidae
Vermicularia fargoi owensi Petuch and Myers, 2014

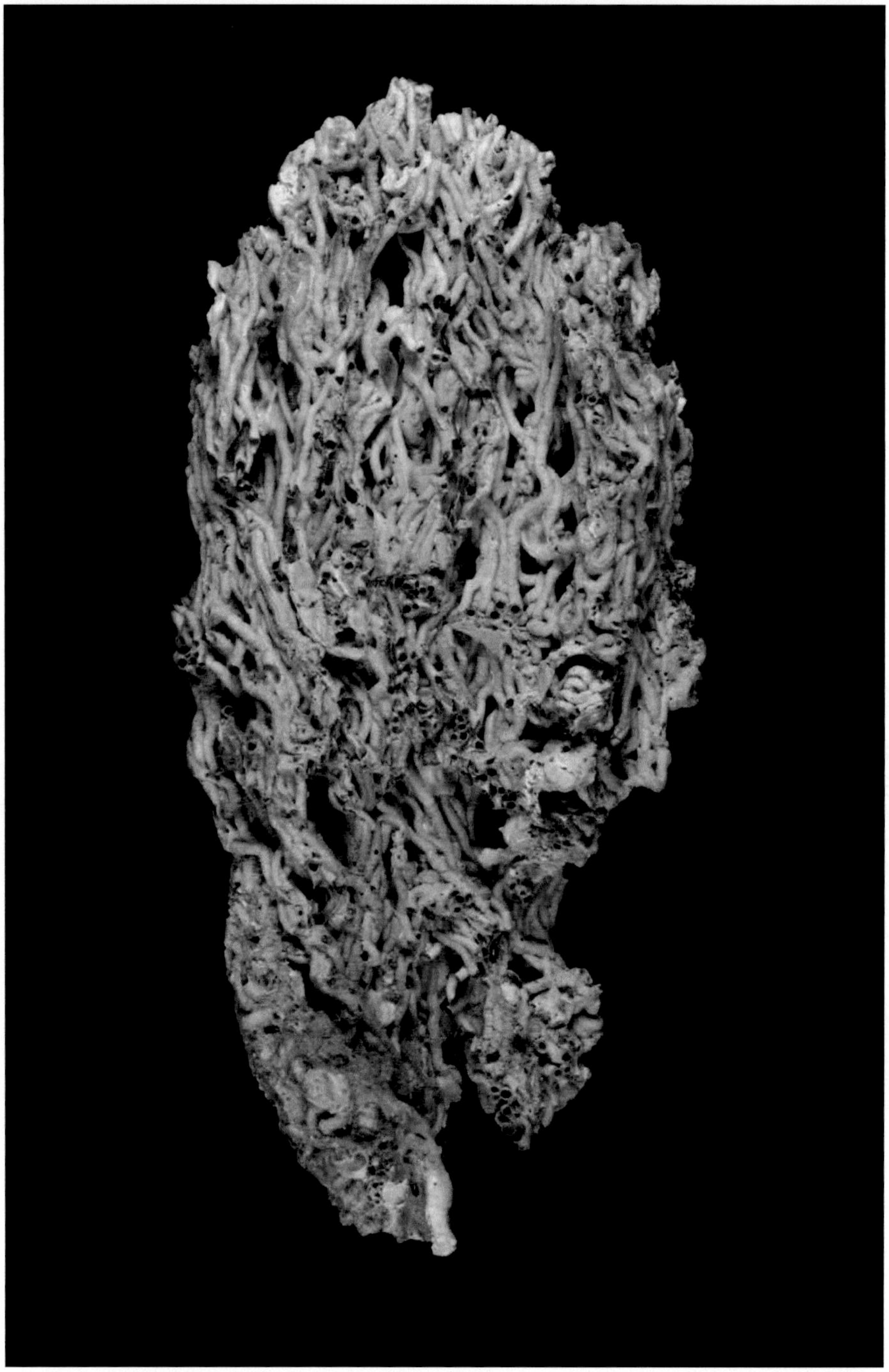

Figure 3.15 An aggregation of the Dark Worm Shell, *Petaloconchus nigricans* (Dall, 1884), length 310 mm, from the vermetoherm off Turtle Key, Ten Thousand Islands, Collier County. These gregarious snails form immense reef-like structures that can be acres in size and produce several growth forms, including flattened sheet-like structures that grow outward into the wave currents (Stromatoverms), flattened concentric ringed structures (Circoverms), and networks of interwoven channels. Being sessile and unable to move, the individual snails feed on plankton and detritus in the water column that is captured on mucus nets that are secreted by the snail's foot.

Figure 3.16 Gastropods of the Vermetoherms.
A, B= *Gemophos tinctus pacei* Petuch and Sargent, 2011, length 31 mm; **C=** *Gemophos tinctus pacei* Petuch and Sargent, 2011, length 27 mm; **D, E=** *Vokesinotus perrugatus* (Conrad, 1846), length 23 mm; **F=** *Vokesinotus perrugatus* (Conrad, 1846), length 15 mm; **G, H=** *Eupleura tampaensis* (Conrad, 1846), length 27 mm; **I=** *Diodora listeri* (d'Orbigny, 1842), length 14.1 mm; **J=** *Lucapinella suffusa* (Reeve, 1850), length 20.5 mm. The muricid genus *Vokesinotus* Petuch, 1988 was named for fossil species and the genus was thought to have become extinct in the mid-Pleistocene. *Vokesinotus* (as *perrugatus*) has now been recognized as having survived along western Florida, from the Ten Thousand Islands north to Cedar Key, and constitutes a "living fossil".

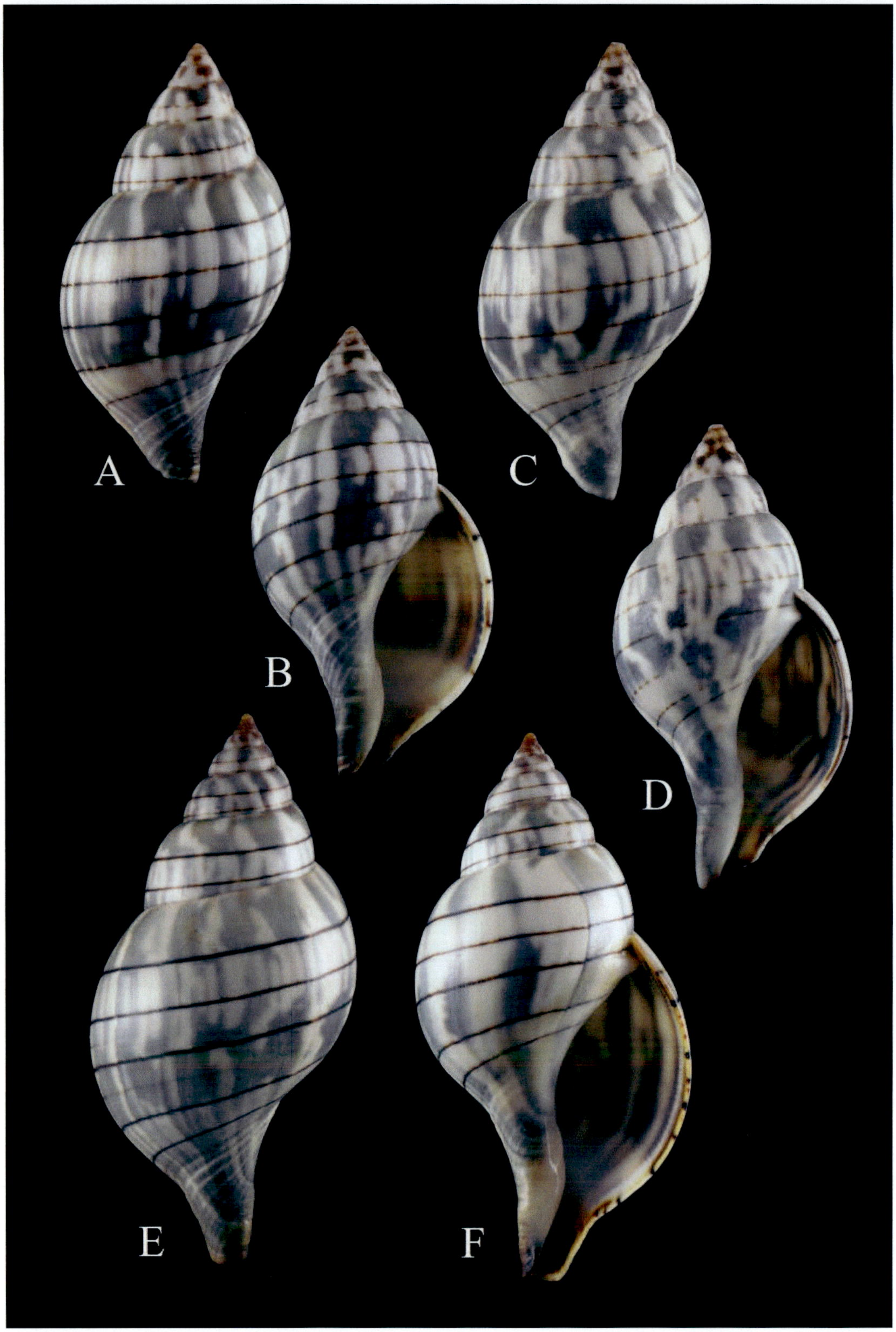

Figure 3.17 The Banded Tulip Shell, *Cinctura hunteria* (Perry, 1811).
A, B= typical specimen, length 62.3 mm; **C, D**= typical specimen, length 63.4 mm; **E, F**= typical specimen, length 78.1 mm.

Figure 3.18 The True Tulip Shell, *Fasciolaria tulipa* (Linnaeus, 1758), showing its color variations.

A, B= typical pale gray and brown specimen, length 118.1 mm; **C, D**= red color form, length 81.6 mm; **E, F**= dark brown and purple color form, length 69.7 mm; **G**= solid dark brown color form, length 56.7 mm; **H**= solid orange-tan color form, length 51.9 mm.

Figure 3.19 Gastropods of the Sponge Banks.
A= *Vermicularia fargoi owensi* Petuch and Myers, 2014, length 130 mm, adult specimen; **B=** *Vermicularia fargoi owensi* Petuch and Myers, 2014, length 105 mm, adult specimen; **C=** *Vermicularia fargoi owensi* Petuch and Myers, 2014, length 94 mm; **D, E=** *Calliostoma tampaense* (Conrad, 1846), typical specimen, length 21 mm; **F, G=** *Calliostoma tampaense* (Conrad, 1846), variant with stepped spire whorls, length 19 mm. **H, I=** *Vermicularia fargoi owensi* Petuch and Myers, 2014, Holotype USNM (Smithsonian Institution) Type No. 1231410, length 23 mm, juvenile.

CHAPTER 4. The Ten Thousand Islands Sand and Mud Flats

Large areas of the shallow sea floors around the Outer Ten Thousand Islands are composed of wide expanses of open sand and mud (Figure 4.1). Since low tides average 1.5 m below the tidal datum throughout the islands, large areas of these soft substrates are exposed for long periods of time. These sand flats and shallow sandy areas frequently manifest themselves as complexes of sand bars that form connections between many of the Outer Ten Thousand Islands (Figure 4.2). Often ephemeral or seasonal in nature, these sand bars and temporary beaches form as the result of heavy wave action during strong winter storms or summer hurricanes. Some outer islands, such as Lumber, Rabbit, Jewel, Camp Lulu, and Pavilion Keys, regularly develop large connecting sand bars that are frequently utilized as camping sites for kayakers traversing the Everglades Kayak Trail. Resulting from the movement of sand during these strong storms, many of the sand bars create entire new bays and lagoons that are cut off, and often partially isolated, from the inter-island channels and main waterways. These partially-enclosed basins then become catchment areas for fine particulate sediments and large amounts of organic matter and frequently house extensive beds of sea grasses (see Chapter 5).

The quartz sand beaches associated with these sand bars are almost always covered with remnants of the marine life that occurs just offshore on the intertidal sand flats and deeper inter-island channels. Most frequently encountered are examples of the rich molluscan fauna that inhabits the shallows around the Outer Ten Thousand Islands, which produce a visually-stunning panoply of brightly-colored and perfectly-preserved shells (Figure 4.3). Similar expanses of accumulated shells are seen from Sanibel Island southward to Marco Island and eventually to Pavilion Key, and typify the entire sandy coastline of the Suwannean Subprovince. Although appearing to contain only assemblages of dead shells and marine life (thanatocoenoses), the sandy beaches of the Outer Ten Thousand Islands also support thriving communities of hardy terrestrial and marine-associated organisms that have evolved a tolerance to the extremes of temperatures seen on the open beaches. One of the most obvious beach-associated invertebrates on the outer islands is the Seaside Dragonlet, *Erythrodiplax berenice* Drury, 1773, a large black dragonfly that can be seen darting about on the sand bars, just above the shoreline accumulations of shells and debris (Figure 4.4). The eggs of the Seaside Dragonlet are laid on algal mats in salt water pools within the interior of the larger islands. Their aquatic larval nymphs also tolerate the high-salinity conditions and metamorphose into adults that feed on biting midges and mosquitos that often swarm the shorelines. For all intents and purposes, *Erythrodiplax berenice* can be considered to be the only marine dragonfly found in the Ten Thousand Islands.

At low tide, the sand flats surrounding the outer islands can be seen to be pitted with thousands of small, circular pools (Figure 4.1). Produced by sting rays feeding at high tide, when they are searching for their small buried crustacean and molluscan prey, these depressions always retain a small amount of sea water and provide temporary shelter for marine organisms stranded at low tide. Along the Outer Ten Thousand Islands, two types of echinoderms are frequently encountered in these small tidal pools; these include asteroids such as the Two-Spined Sand Star (*Astropecten duplicatus* Gray, 1840 (Figure 4.5) and ophiuroids such as the Short-Spined Brittle Star (*Ophioderma brevispinum* (Say, 1825) (Figure 4.6). These two echinoderms share the sand flats and pools with the overwhelmingly-abundant Least Horn Shell, *Lampanella minima* (Gmelin, 1791), which often forms densely-packed aggregations that cover acres of exposed sand flats (Figure 4.7). These small batillariid snails are, without doubt, the most abundant mollusk in the Ten Thousand Islands and they can be seen crawling on the wet sand by the millions. Here, they feed on algal and bacterial films that grow on the sand surface and, in turn, serve as a food resource for numerous birds, crabs, fish, and large predatory gastropods. Juvenile specimens of the largest American gastropod, the Horse Conch (*Triplofusus giganteus* (Kiener, 1840)), are frequently encountered on exposed sand flats, feeding on different species of Ceriths (Horn Shells) (such as the related *Cerithium muscarum* shown in Figure 4.9). Other predators of the exposed sand and mud flats of the Ten Thousand Islands include several species of herons and egrets, most notably the Yellow-Crowned Night Heron, *Nyctanassa violacea* (Linnaeus, 1758) (Figure 4.10). This conspicuous shore bird feeds primarily on small crabs and shrimp, which it captures and crushes with its specially-adapted thick, heavy bill.

The intertidal sand flats also house a number of other large carnivorous invertebrates, with the Horseshoe Crab, *Limulus polyphemus* (Linnaeus, 1758), being one of the more abundant and prominent species (Figure 4.11). These "living fossils" have remained unchanged morphologically since the Silurian Period and they, and two other species from the Red Sea and China, are the only living relatives of the Paleozoic Sea Scorpions. Crawling over the sand flats at low tide, Horseshoe Crabs feed on small mollusks, polychaete worms, carrion, and algae. Juvenile *Limulus* frequently fall victim to shore bird predation, but if they survive to adulthood, they can live for up to 20 years. The carcasses of freshly dead, very old individuals are commonly found during the summer months and these are a major food resource for numerous carrion-feeding gastropods such as *Melongena (Rexmela) corona* (Figure 4.12) and several species of nassariid Mud Snails. Another prominent sand flat predator is the large Left-Handed Whelk, *Sinistrofulgur sinistrum* (Hollister, 1958) (Figure 4.13), which feeds exclusively on large clams that are buried in the sand substrate. Upon locating its bivalve prey, the Left-Handed Whelk envelops the clam with its foot and then begins to methodically and rhythmically pound the edge of its shell lip against the junction of the two valves of the clam (commissure); in essence acting as a chisel blade that can chip away a small hole. Once the hole is formed, the whelk then inserts its tooth-filled, worm-like proboscis through the opening and rasps away pieces of the unprotected tissue. Left-Handed Whelks are abundant on the Outer Ten Thousand Islands sand flats and mating pairs can be seen on many exposed areas during the late spring (Figure 4.14). After mating, the female whelk lays a long string of egg capsules within the deeper channel areas between the islands and these often wash ashore after heavy spring storms (Figure 4.15). Another molluscan predator of the sand flats is the Lettered Olive Shell, *Americoliva sayana sarasotaensis* (Petuch and Sargent, 1986) (Figure 4.16), which feeds on smaller mollusks, including juvenile Left-Handed Whelks, and also carrion.

Besides the large Left-Handed Whelks, the primary molluscan predators of the sand and mud flats are the Moon Snails of the family Naticidae (examples shown here in Figure 4.22). At least four species occur throughout the Ten Thousand Islands and these feed on both bivalves and gastropods and often are cannibalistic, feeding on smaller specimens of their own species. Unlike the whelks, which use the edge of their apertural lip to chip gaps between the valves of their bivalve prey, the moon snails use a combination of acid and their rows of sharp apatite radular teeth to drill perfectly circular holes in the shells of their victims. Once the beveled hole is drilled through the shell, the moon snail inserts its long, tooth-filled proboscis and rasps out pieces of its living victim. The extremely large numbers of small drilled clam shells found on the Ten Thousand Islands beaches attest to the efficiency and ferocity of the naticid predatory gastropods. The two largest moon snail species found on the sand flats, *Neverita (Glossaulax) delessertiana* (Recluz, 1843) and *Neverita (Glossaulax) duplicata campechiensis* (Recluz, 1843) (both shown in Figure 4.22), feed on larger bivalves such as juvenile individuals of the Campeche Venus Clam, *Mercenaria campechiensis* (Gmelin, 1791), Brown's Venus Clam, *Mercenaria browni* Petuch and Berschauer, 2019, the Sunray Venus, *Macrocallista nimbosa* (Lightfoot, 1786), the Ponderous Ark Shell, *Noetia ponderosa* (Say, 1822), and the Egmont Cockle Shell, *Trachycardium egmontianum* (Shuttleworth, 1856). Smaller naticid species, such as the Colorful Atlantic Natica, *Naticarius canrena* (Linnaeus, 1758), and Vera's Moon Snail, *Naticarius verae* Rehder, 1847 (both shown here in Figure 4.22), feed primarily on smaller sand flat bivalves and the abundant *Lampanella minima* horn snails. Vera's Moon Snail is one of the special Ten Thousand Islands species, as it is endemic to the archipelago and is found nowhere else on Earth (Rehder's holotype was collected on Marco Island).

The predatory busyconid and naticid gastropods share their sand and mud flat environment with a host of other conspicuous mollusks, including large aggregations of the Florida Fighting Conch, *Strombus alatus* (Gmelin, 1791) (Figure 4.18). These small, gregarious conch shells congregate in large numbers on the sand flats at high tide, grazing on algal films that cover the exposed sand surface. The abundant organic material that derives from the Everglades river effluent mixes with sediments on the sand flats and supports a large fauna of detritivore and filter-feeding polychaete worms. These worms, in turn, serve as the sole food resource for a fauna of small vermivorous gastropods. Principal among these are Stearns' Cone Shell, *Jaspidiconus stearnsi* (Conrad, 1869), the Auger Shells, *Neoterebra vinosa* (Dall, 1889), and *Neoterebra dislocata* (Say, 1822) (all shown here in Figure 4.23). Several small scavenger and carrion-feeding gastropods occur in abundance along with the vermivorous species, including Swearingen's Mud Snail, *Uzita swearingeni* Petuch and Myers, 2014, the Common Mud Snail, *Phrontis vibex* (Say, 1822), and the Common Atlantic Marginella, *Prunum apicinum* (Menke, 1828) (all shown here in Figure 4.23).

The mud and sandy-mud flats, especially those within quiet, enclosed lagoons, are the preferred habitats for deeply-burrowing pholadid bivalves. The largest and most beautiful of these, *Cyrtopleura costata* (Linnaeus, 1758), also known as the "Angel Wing" (Figure 4.27), is highly sought-after by shell collectors. Adding to the mystery and desirability of this large pholad is the fact that it is also the most deeply-burrowing bivalve found on the mud flats, sometimes digging to depths of 1 m below the sea floor. Being safely sequestered well below the mud surface, the Angel Wing extends its very long siphons upward into the water column and pumps plankton-rich and oxygenated water down to the body of the animal. A smaller and far more fragile species of pholad, the Mud Pholad, *Barnea truncata* (Say, 1822) (Figure 4.28), occurs together with *Cyrtopleura costata* on the enclosed mud flats but does not dig as deeply and is therefore much more easily dislodged during storms and cast up onto the beach. Single valves of *Barnea truncata* are common on beaches along Pavilion, Rabbit, Turtle, and Jewell Keys and frequently exhibit a distinctive pale lavender or pink shell color.

Pictorial Overview of Intertidal Sand and Mud Flats and Associated Organisms

The following photographs show some of the most conspicuous organisms that are present on the intertidal sand and mud flats of the Ten Thousand Islands.

Figure 4.1 View of a sand flat, during an extremely low Spring Tide, off Turtle Key, Ten Thousand Islands, Collier County. The numerous small tide pools, most formed by sting rays feeding during high tide, act as temporary refuges for many marine organisms that were trapped by the receding tide.

Figure 4.2 View of Rabbit Key from Lumber Key, Ten Thousand Islands, Monroe County, showing the large sand peninsula that connects the two islands. The large, shallow bay to the left of the sand peninsula houses a classic muddy sea floor environment, much of which is exposed at low tide.

View of Lumber Key from Rabbit Key.

Figure 4.3 Close-up view of an untouched sand beach on Rabbit Key, Ten Thousand Islands, showing both the abundance of large shells and the high diversity of species that typically wash up onto the islands. Here, in this one small patch of sand, at least nine species of mollusks can be seen, including gastropods such as the Tulip Shells *Fasciolaria tulipa* and *Cinctura hunteria*, the Florida Keys Pear Whelk *Fulguropsis keysensis*, the Fig Shell *Ficus papyratia*, the Slipper Shell *Crepidula fornicata*, the Moon Snail *Neverita (Glossaulax) delessertiana*, and the Worm Shell *Petaloconchus nigricans*, and bivalves such as the Spiny Cockle *Trachycardium egmontianum* and the Virginia Oyster *Crassostrea virginica*.

Figure 4.4 Close-up view of a male Seaside Dragonlet, *Erythrodiplax berenice* Drury, 1773, resting on an oyster shell on the beach at Pavilion Key, Ten Thousand Islands, Monroe County. This large dragonfly is one of the few insects that can tolerate salt water conditions and is commonly seen darting about, in search of prey, along the water line on most beaches of the Outer Ten Thousand Islands. Photo courtesy of Morris A. Foster.

Figure 4.5 Close-up view of a pair of the Two-Spined Sand Star, *Astropecten duplicatus* Gray, 1840, in a tide pool on the sand flats off Turtle Key, Ten Thousand Islands, Collier County. The sand stars share their temporary home during low tide with several small red sponges.

Figure 4.6 Close-up view of the Short-Spined Brittle Star, *Ophioderma brevispinum* (Say, 1825), in a tide pool on the sand flats off Turtle Key, Ten Thousand Islands, Collier County.

Figure 4.7 View of a sand flat, at low tide, on Turtle Key, Ten Thousand Islands, Collier County, showing an immense aggregation of the Least Horn Shell, *Lampanella minima* (Gmelin, 1791) (Family Batillariidae). These small snails feed on algal and bacterial films that grow on the sand flats and often occur in swarm-like masses composed of millions of individuals, making them the most abundant mollusk in the Ten Thousand Islands.

Figure 4.8 Close-up view of numerous individuals of the Least Horn Shell, *Lampanella minima* (Gmelin, 1791), covering an algal mat off Turtle Key, Ten Thousand Islands.

Figure 4.9 Close-up view of a bright orange juvenile of the Horse Conch, *Triplofusus giganteus* (Kiener, 1840), on a sand flat off Turtle Key, Ten Thousand Islands. Here, the voracious baby Horse Conch is seen devouring a Fly-Specked Horn Shell, *Cerithium muscarum* Say, 1822, and has inserted its proboscis into the aperture of its hapless victim. The leaked body fluids of the dying horn shell have attracted a scavenger Mud Snail, *Phrontis vibex* (Say, 1822), which has extended its long siphon in hopes of garnering a few left-over scraps of horn shell flesh.

Figure 4.10 A Yellow-Crowned Night Heron, *Nyctanassa violacea* (Linnaeus, 1758), hunts for crustaceans on a mud flat, at low tide, off Chokoloskee Island, Collier County. The thick, heavy bill of the Night Heron is an adaptation for crushing the shells of its crab and shrimp prey. Photo courtesy of Michael Bruggeman.

Figure 4.11 A pair of Horseshoe Crabs, *Limulus polyphemus* (Linnaeus, 1758), crawling on a sand flat, at low tide, off Turtle Key, Ten Thousand Islands. These shallow water animals are actually not crabs at all, but are distantly related to spiders and to the extinct Sea Scorpions of the Silurian Period and can be considered "living fossils". Photo courtesy of Michael Bruggeman.

Figure 4.12 Close-up view of a dead and decomposing Horseshoe Crab being scavenged by six Crown Conchs, *Melongena (Rexmela) corona*, on a mud flat off Rabbit Key, Ten Thousand Islands. Although primarily predators on oysters, clams, and other gastropods, Crown Conchs will also feed on carrion, particularly dead crustaceans and Horseshoe Crabs, when available.

Figure 4.13 A specimen of the Left-Handed Whelk, *Sinistrofulgur sinistrum* (Hollister, 1958), on a sand flat off Turtle Key, Ten Thousand Islands. These large and common gastropod mollusks are voracious predators on clams. The whelks locate buried clams by smell and then use the sharp lip edge of their shells to chip a hole along the juncture of the clam's two shells. Once a small break is produced, the whelk inserts its long, tooth-filled proboscis through the hole and rasps out chunks of the living flesh of the clam victim. A small green Swearingens' Mud Snail, *Uzita swearingeni* Petuch and Myers, 2014, occupied by a hermit crab, is seen crawling in the sand directly below the whelk.

Figure 4.14 Two Left-Handed Whelks, *Sinistrofulgur sinistrum* (Hollister, 1958), mating on a sand flat off Turtle Key, Ten Thousand Islands. After mating is complete, the female will lay a long egg string containing several hundred individual egg capsules, each containing 20-30 larval snails.

Figure 4.15 Close-up view of a string of egg capsules that was laid by a Left-Handed Whelk, *Sinistrofulgur sinistrum* (Hollister, 1958). The mother snail buries the end of the egg string into the sand in deeper water offshore, anchoring it against strong currents and storm surge. During exceptionally strong storm turbulence, however, the egg strings are ripped from the sea floor and wash ashore, like this example from the beach at Turtle Key, Ten Thousand Islands. Photo courtesy of Michael Bruggeman.

Figure 4.16 Close-up view of a living Western Florida Olive Shell, *Americoliva sayana sarasotaensis* (Petuch and Sargent, 1986), crawling on a sand flat, at low tide, off Pavilion Key, Ten Thousand Islands, Monroe County. True *Americoliva sayana* ranges from North Carolina south to Palm Beach County, Florida and is not found in the Florida Keys or the area extending from Pompano Beach to Miami. The Florida west coast *A. sayana sarasotaensis* is genetically isolated from the east coast populations. Another subspecies, *A. sayana texana* (Petuch and Sargent, 1986) is restricted to the Texas and eastern Mexico coast. Photo courtesy of Robert L. Eason, Sr.

Iconography of Sand and Mud-Associated Mollusks

The gastropod and bivalve mollusks illustrated here are all classic inhabitants of the Ten Thousand Islands intertidal sand and mud flats. Some of the more important ecological index species include:

Gastropoda
Neritidae
Vitta usnea (Roding, 1798)
Batillariidae
Lampanella minima (Gmelin, 1791)
Strombidae
Strombus alatus (Gmelin, 1791)
Naticidae
Naticarius canrena (Linnaeus, 1758)
Naticarius verae Rehder, 1947
Neverita (Glossaulax) delessertiana (Recluz, 1843)
Neverita (Glossaulax) duplicata campechiensis (Recluz, 1843)
Fasciolariidae
Triplofusus giganteus (Kiener,1840)
Triplofusus giganteus form *reevei* (Jonas, 1850)
Busyconidae
Sinistrofulgur sinistrum (Hollister, 1958)
Nassariidae
Phrontis vibex (Say, 1822)
Uzita swearingeni Petuch and Myers, 2014
Marginellidae
Prunum apicinum (Menke, 1828)
Olividae
Americoliva sayana sarasotaensis (Petuch and Sargent, 1986)
Conidae
Jaspidiconus acutimarginatus (Sowerby II, 1866)
Jaspidiconus stearnsi (Conrad, 1869)
Terebridae
Neoterebra cf. *dislocata* (Say, 1822)
Neoterebra vinosa (Dall, 1889)

Bivalvia
Arcidae
Anadara floridana (Conrad, 1869)
Anadara transversa (Say, 1822)
Noetiidae
Noetia ponderosa (Say, 1822)

Cardiidae
Trachycardium egmontianum (Shuttleworth, 1856)

Veneridae
Macrocallista nimbosa (Lightfoot, 1786)
Mercenaria browni Petuch and Berschauer, 2019
Mecenaria campechiensis (Gmelin, 1791)
Mecenaria mecenaria notata (Say, 1822) (introduced from North Carolina)
Tellinidae
Eurytellina lineata (Turton, 1819)
Macoma tenta (Say, 1834)
Tampaella tampaensis (Conrad, 1866)
Mactridae
Mactrotoma fragilis (Gmelin, 1791)
Pholadidae
Barnea truncata Say, 1822
Cyrtopleura costata (Linnaeus, 1758)

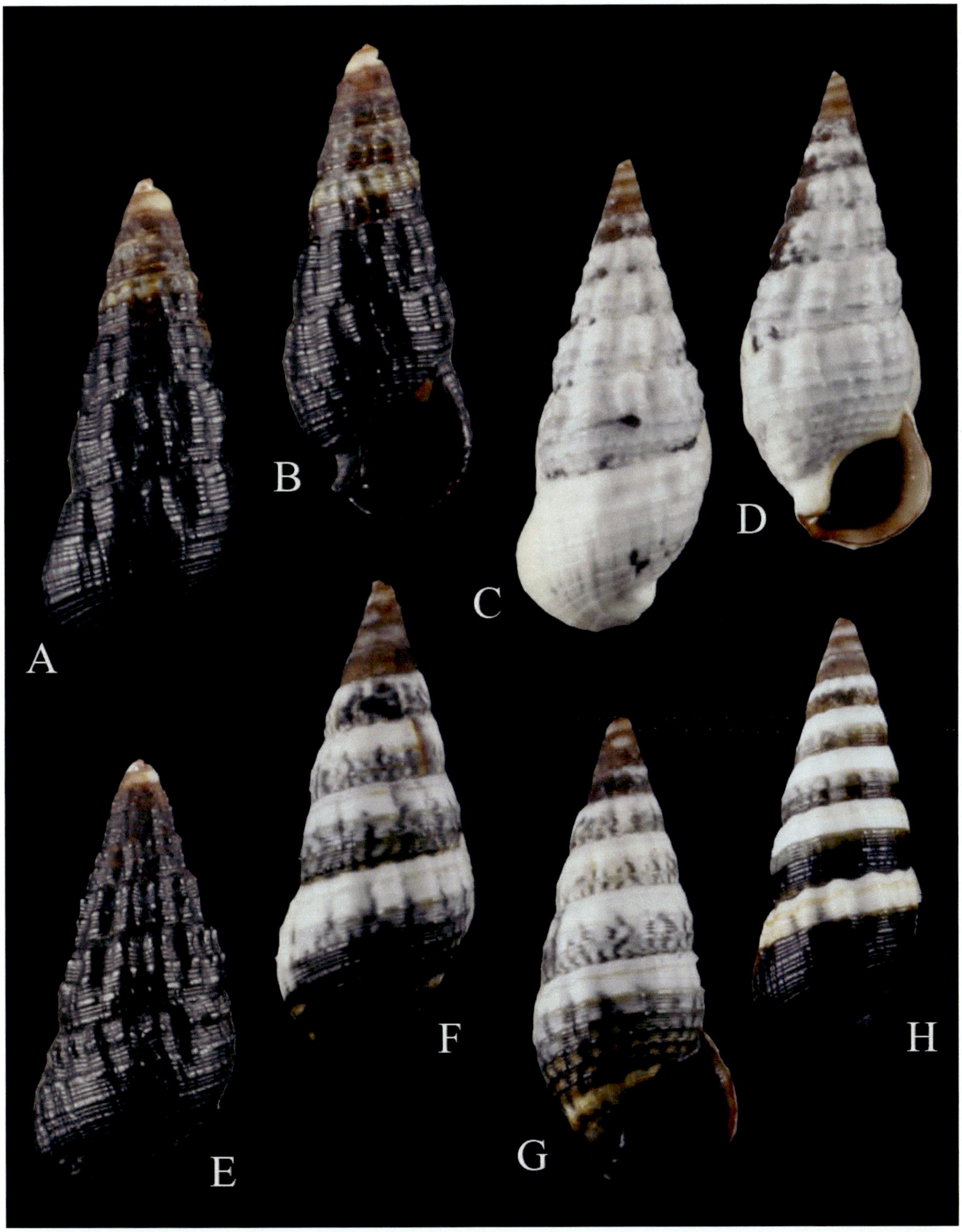

Figure 4.17 The Least Horn Shell, *Lampanella minima* (Gmelin, 1791), showing its color variations.
A, B= typical black specimen, length 14 mm; **C, D**= pale gray color form, length 16 mm; **E**= typical black specimen, length 14 mm; **F, G**= color form with a wide, mottled band around the shoulder area, length 15 mm; **H**= color form with a narrow white band around the shoulder area, length 13 mm.

Figure 4.18 The Florida Fighting Conch, *Strombus alatus* (Gmelin, 1791).
A, B= typical specimen, length 96 mm; **C, D**= specimen with a purple aperture and pale shell color, length 93 mm; **E**= typical dark-colored specimen, length 95 mm.

Figure 4.19 The Horse Conch, *Triplofusus giganteus* (Kiener, 1840).
A, B= typical specimen, length 271 mm.

Figure 4.20 The Knobless Horse Conch, *Triplofusus giganteus* form *reevei* (Jonas, 1850).
A, B= typical specimen, length 278 mm.

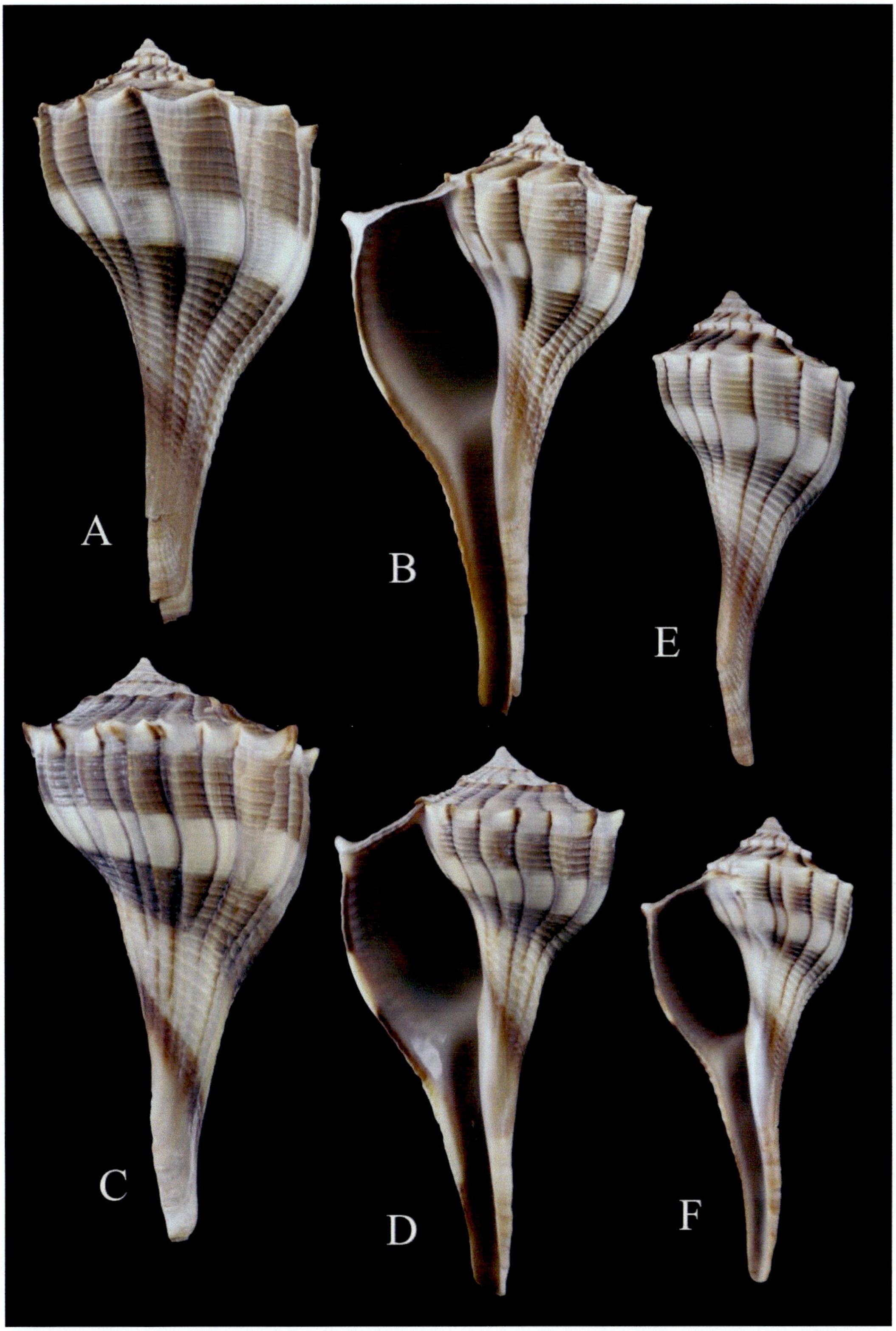

Figure 4.21 The Left-Handed Whelk, *Sinistrofulgur sinistrum* (Hollister, 1958).
A, B= typical specimen, length 105 mm; **C, D**= typical specimen, length 97 mm; **E, F**= typical specimen, length 93 mm.

Figure 4.22 Moon Snails (Family Naticidae) of the Ten Thousand Islands Sand Flats. **A, B=** *Naticarius canrena* (Linnaeus, 1758), length 32 mm; **C, D=** *Naticarius verae* Rehder, 1947, length 28 mm; **E, F=** *Naticarius verae* Rehder, 1947, length 27 mm; **G, H=** *Neverita (Glossaulax) delessertiana* (Recluz, 1843), length 33 mm; **I, J=** *Neverita (Glossaulax) duplicata campechiensis* (Recluz, 1843), length 18 mm. The brown-banded *Naticarius verae* is endemic to the Ten Thousand Islands archipelago.

Figure 4.23 Gastropods of the Ten Thousand Islands Sand Flats.
A= *Prunum apicinum* (Menke, 1828), length 11 mm; **B, C=** *Phrontis vibex* (Say, 1822), length 12 mm; **D=** *Jaspidiconus stearnsi* (Conrad, 1869), lavender color form, length 21 mm; **E, F=** *Jaspidiconus stearnsi* (Conrad, 1869), typical specimen, length 23 mm; **G=** *Jaspidiconus acutimarginatus* (Sowerby II, 1866), length 20 mm (most frequently encountered in the Florida Keys; rare in the Ten Thousand Islands); **H=** *Neoterebra dislocata* (Say, 1822), length 32 mm; **I=** *Neoterebra vinosa* (Dall, 1889), length 14 mm; **J, K=** *Uzita swearingeni* Petuch and Myers, 2014, length 9 mm.

Figure 4.24 ***Americoliva sayana sarasotaensis*** **Petuch and Sargent, 1986**
A, B= typical specimen, length 47 mm; **C, D**= pale-colored specimen, length 44 mm; **E, F**= golden color form *citrina*, length 32 mm; **G, H**= dark color form, length 50 mm.

Figure 4.25 Venerid Bivalves of the Ten Thousand Islands Sand Flats.
A= *Macrocallista nimbosa* (Lightfoot, 1786), length 109 mm; **B**= *Macrocallista nimbosa* (Lightfoot, 1786), dark color form, length 88 mm; **C**= *Macrocallista nimbosa* (Lightfoot, 1786), pale color form, length 89 mm; **D**= *Mercenaria campechiensis* (Gmelin, 1791), left valve, length 105 mm; **E**= *Mercenaria campechiensis* (Gmelin, 1791), left valve, length 92 mm.

Figure 4.26 Bivalves of the Ten Thousand Islands Sand Flats.
A, B= *Trachycardium egmontianum* (Shuttleworth, 1856), exterior and interior of the same valve, length 48 mm; **C=** *Mactrotoma fragilis* (Gmelin, 1791), length 35 mm; **D=** *Macoma tenta* (Say, 1834), length 35 mm; **E=** *Tampaella tampaensis* (Conrad, 1866), length 31 mm; **F, G=** *Eurytellina lineata* (Turton, 1819), length 28 mm; **H=** *Noetia ponderosa* (Say, 1822), length 45 mm; **I=** *Anadara floridana* (Conrad, 1869), length 76 mm; **J=** *Anadara transversa* (Say, 1822), length 38 mm.

Figure 4.27 Brown's Venus Clam, *Mercenaria browni* Petuch and Berschauer, 2019. **A=** left valve of a typical specimen, length 91 mm; **B, C=** exterior and interior of a frilly, heavily-sculptured juvenile specimen, length 49 mm; **D=** left valve of a juvenile specimen, length 41 mm. Brown's Venus Clam is endemic to the Ten Thousand Islands, where it lives in estuaries at the mouths of the numerous rivers that drain the Everglades. It is most common from Gomez Key south to Pavilion Key and is frequently encountered along the mouth of the Rabbit Key Pass near Turtle Key.

Figure 4.28 The Angel Wing Pholad, *Cyrtopleura costata* (Linnaeus, 1758).
A, B= typical specimen, length 153 mm. This fragile bivalve lives deeply buried in muddy sand, often to depths of 1 m below the sea floor surface. Because of its great beauty, the Angel Wing is highly sought after by shell collectors and a large perfect pair is considered to be a major collector's item.

Figure 4.29 Close-up view of a single valve of a 38 mm Mud Pholad, *Barnea truncata* (Say, 1822), washed up onto the beach along Rabbit Key, Ten Thousand Islands, Monroe County, Florida. The specimens from mud flats around Rabbit Key are frequently colored a pale lavender or lavender-pink.

View of an exposed Shoal Grass (*Halodule wrightii* Ascherson, 1868) bed at low tide along Rabbit Key.

CHAPTER 5. The Ten Thousand Islands Sea Grass Beds

In the shallow protected bays and lagoons of the Outer Ten Thousand Islands, large amounts of mud and organic matter can accumulate and produce the nutrient-rich substrate that is needed to support the development of sea grass beds (Figure 5.1). Three types of sea grasses are most often seen in these lagoons and bays, depending on the depth of the water and the stability of the salinity, turbidity, and water chemistry. These include: **Turtle Grass** (*Thalassia testudinum* Banks and Solander *in* Koenig, 1805); **Manatee Grass** (*Syringodium filiforme* Kutzing *in* Hohenacker, 1860); and **Shoal Grass** (*Halodule wrightii* Ascherson, 1868). Of these three classic western Atlantic sea grasses, Shoal Grass (*Halodule*) is the most abundant type in the shallow muddy lagoons and often produces densely-packed meadows that cover acres of sea floor (Figure 5.2). As the most physiologically-tolerant of the three main sea grasses, *Halodule* can thrive in areas that would kill the other types. Shoal Grass has evolved the abilities to withstand subaerial conditions for long periods of time, tolerate wildly-fluctuating salinities, and experience high levels of suspended sediments and extreme turbidity. Since the islands have a wide tidal range with extreme spring tides, experience heavy tropical rainfall during the summer, and are near river mouths that discharge mud-sized particles, Shoal Grass is uniquely suited to be the predominant sea grass throughout the archipelago. Turtle Grass and Manatee Grass, which both prefer a constant higher salinity and clearer, less turbid water, grow only in deeper water areas offshore of the northern and extreme southern islands and are not found in most of the shallow bays, lagoons, and inlets.

The Shoal Grass beds, which are exposed at low tide and often cover acres of sea floor, harbor a large number of conspicuous marine invertebrates. Primary among these is the Spiny Sea Star, *Echinaster sentis* (Say, 1825) (Figure 5.3), which is a predator on small bivalves and crustaceans. Competing with *Echinaster* for bivalve prey is the Western Florida Pear Whelk, *Fulguropsis pyruloides* (Say, 1822) (Figures 5.4 and 5.7), which is abundant in *Halodule* beds along the entire archipelago. The Ten Thousand Islands populations of this small busyconid gastropod represent the southernmost end of the range of the species, which extends from Pavilion Key northward to Mobile Bay, Alabama. Living along with *Fulguropsis pyruloides* in the Shoal Grass beds is another small whelk that is closely related: the Florida Keys Pear Whelk, *Fulguropsis keysensis* Petuch, 2013 (Figure 5.8). This species is similar to the Western Florida Pear Whelk in both shape and size, but differs in being a much more heavily-sculptured shell, ornamented with numerous strong spiral ribs instead of having a smooth, silky shell texture. The Ten Thousand Islands population of this second small busyconid gastropod represents the northernmost extension of the range of the species, which extends from Biscayne Bay, across the entire Florida Keys island chain out to the Dry Tortugas, and northward to Marco Island. The Outer Ten Thousand Islands is the only place where the ranges of the two similar species overlap and where both whelks occur together in the same ecosystem. Here, in the Shoal Grass beds, they compete for the same bivalve prey, which generally consists of thin-shelled, shallowly-burrowing species such as the

Morton's Egg Cockle, *Laevicardium mortoni* (Conrad, 1813), the Dwarf Venus Clam, *Anomalocardia cuneimeris* (Conrad, 1846) (both Figure 5.12), and the tellinid *Angulus sybariticus* (Dall, 1881) (Figure 5.13). These small bivalves, and several others shown here in Figures 4.12 and 4.13, often live buried within the dense rhizome root mats of the Shoal Grass beds; this habitat offers a protected area where thin-shelled species can avoid predation from voracious busycon whelks.

The Sea Grass beds also provide the habitat for a rich and diverse fauna of smaller gastropods, many of which occur in large aggregations. Living directly on the Shoal Grass blades, and feeding on encrusting algal growths, the Florida Button Shell, *Modulus floridanus* (Conrad, 1869) (Figure 5.5), is often present in large numbers. The shell of his small algivorous gastropod, itself, provides the substrate for a tiny commensal slipper shell, *Crepidula ustulatulina* (Collin, 2002) (Figure 5.5). This small crepidulid gastropod fixes itself to the spire of the slipper shell and never moves, feeding by filtering sea water in much the same fashion as bivalve mollusks. The small Fly-Speck Cerith Shell, *Cerithium muscarum* Say, 1822, lives along with *Modulus floridanus* on the *Halodule* blades and grazes on epiphytic algal growth. The small sponges, hydroids, and bryozoan ectoprocts that encrust the grass blades also serve as food resources for small grazing carnivores such as the Half-Ribbed Dove Shell, *Coastanachis semiplicata* (Stearns, 1873) (Figure 5.5G, H), Taylor's Dove Shell, *Falsuszafrona taylorae* (Petuch, 1987) (Figure 5.5K, L), the Western Bubble Shell, *Bulla occidentalis* A. Adams, 1850 and Taylor's Bubble Shell, *Haminoea taylorae* Petuch, 1987, both slug-like snails that feed on small bivalves and gastropods (both Figure 5.5). The large muricid gastropods, *Vokesimurex rubidus* (F.C. Baker, 1897), *Phyllonotus pomum* (Gmelin, 1791), and *Chicoreus dilectus* (A. Adams, 1855) are often found partially buried in the rhizome root mat, where they search for small bivalve prey. Once encountered, the muricids use their file-like radular teeth to drill a hole in the shell of their victim, much like moon shells of the family Naticidae. Unlike the perfectly-round and beveled drill holes of the naticids, the holes produced by muricid shells are uneven in shape, with ragged edges along the inside.

Among the most conspicuous and interesting gastropods of the sea grass beds are two species of large cone shells: the Atlantic Alphabet Cone, *Lindaconus atlanticus* (Clench, 1942) (Figure 5.9); and the Florida Cone, *Gradiconus floridanus* (Gabb, 1869). Both species, belonging to different genera, are voracious marine worm eaters, which search for the buried polychaete prey among the *Halodule* rhizomes and dense root mats. The polychaete sea worm prey is located by smell (chemoreceptors), and once uncovered, is immediately harpooned by a large, hollow, arrow-shaped tooth. Once the worm is bitten, powerful venom and digestive enzymes are injected into the prey, both paralyzing and predigesting it before being swallowed. Of these two cone shells, the Florida Cone has a somewhat convoluted nomenclatural history, leading to some recent confusion. The name *Conus floridanus* has been a well-known taxon that was used in both scientific and popular literature for over 150 years (*i.e.* R.T. Abbott's "American Seashells", 1974), and Gabb's holotype of *floridanus* is an example of the typical yellow color form found along western Florida. In the past twenty years, however, Gabb's species name has been incorrectly synonymized by many workers with an older name, "*Conus anabathrum* Crosse, 1865"; that taxon is now known to represent a different species from the Eastern Pacific Panamic Province (Berschauer, 2022) and is not the same as the western Florida shell. This abundant and well-known western Floridian species occurs in two distinct color forms: the golden-yellow variant which represents the classic *floridanus* (Gabb, 1869) (Figure 5.10); and the dark-colored and highly patterned variant that was named *floridensis* (Sowerby II, 1870) (Figure 5.11). Both color forms occur together in any given population, with the *floridensis* form being the prevalent variety found along the central Outer Ten Thousand Islands.

The individual Shoal Grass plants, themselves, also offer a habitat for specialized bivalves, especially mussels of the family Mytilidae. A classic example of these is the extremely fragile and thin-shelled Paper Mussel, *Arcuatula papyria* (Conrad, 1846) (Figure 4.12), which lives attached to the stems and blades of single plants. As is typical of its family, the Paper Mussel attaches itself to the long, thin grass blades by strong, flexible byssal threads. Another member of the Mytilidae, the much larger Tulip Mussel, *Modiolus americanus* (Leach, 1815) (Figure 5.13), attaches to the base of the individual plants and frequently is overgrown by portions of the dense root mat. In larger Shoal Grass beds, with thick and well-developed growth (such as seen here in Figure 5.2), Taylor's Bay Scallop, *Argopecten irradians taylorae* Petuch, 1987 (Figure 5.14), often occurs in large aggregations of thousands of individuals. These free-living pectinid bivalves lie on their side, right valve down, and swim by jet propulsion. When startled, the entire shoal of scallops jumps upward into the water column or scoots laterally through the dense sea grass growth. The individual scallops swim by rapidly opening and closing their shells and squirting jets of water out through the two openings along the hinge line. Taylor's Bay Scallop was commercially fished within the boundaries of Everglades National Park up until the early 1980s, but it is now protected and the commercial scallop boats have moved elsewhere up the western coast of Florida. Also living buried deeply within the sea grass rhizome root mats and sand, throughout the Outer Ten Thousand Islands, are three species of large Pen Shells (family Pinnidae), including *Atrina rigida* (Lightfoot, 1786), *Atrina seminuda* (Lamarck, 1786), and *Atrina serrata* (Sowerby I, 1825) (all shown here in Figure 5.15). These distant relatives of the scallops anchor themselves in the mud and sand by long, thick mats of byssal fibers and only the top edge of their triangular shells protrude above the sediments.

Pictorial Overview of Sea Grass Environments and Associated Organisms

The following photographs show typical sea grass environments and some of the more conspicuous sea grass organisms that are present along the Ten Thousand Islands.

Figure 5.1 View of the western side of Jewell Key, Ten Thousand Islands, Collier County, showing small patches of sea grasses growing in the shallow adjacent bay.

Figure 5.2 Close-up view of a sea grass meadow composed of Shoal Grass, *Halodule wrightii* Ascherson, 1868, growing in shallow water along the northern side of Rabbit Key, Ten Thousand Islands, Monroe County. Shoal Grass is the most abundant sea grass in the sandy intertidal areas off many of the islands.

Figure 5.3 Close-up view of the Spiny Sea Star, *Echinaster sentus* (Say, 1825), a common asteroid echinoderm found in most of the Shoal Grass beds around the Ten Thousand Islands.

Figure 5.4 Close-up view of the Western Florida Pear Whelk, *Fulguropsis pyruloides* (Say, 1822), partially buried in sand in a Shoal Grass bed off Rabbit Key, Ten Thousand Islands.

Iconography of Sea Grass-Associated Mollusks

The gastropod and bivalve mollusks illustrated here are all classic inhabitants of the Ten Thousand Islands sea grass beds. The more important ecological index taxa include:

Gastropoda
Crepidulidae
Crepidula ustulatulina (Collin, 2002)
Cerithiidae
Cerithium muscarum (Say, 1822)
Modulidae
Modulus floridanus (Conrad, 1869)
Muricidae
Chicoreus dilectus (A. Adams, 1855)
Phyllonotus pomum (Gmelin, 1791)
Vokesimurex rubidus (F.C. Baker, 1897)
Busyconidae
Fulguropsis keysensis Petuch, 2013
Fulguropsis pyruloides (Say, 1822)
Columbellidae
Costanachis semiplicata (Stearns, 1873)
Falsuszafrona taylorae (Petuch, 1987)
Conidae
Gradiconus floridanus (Gabb, 1869)
Gradiconus floridanus (Gabb, 1869) form *floridensis* (Sowerby II, 1870)
Lindaconus atlanticus (Clench, 1942)
Lindaconus atlanticus (Clench, 1942) form *aureofasciatus* (Rehder & Abbott, 1951)
Bullidae
Bulla occidentalis A. Adams, 1850
Haminoeidae
Haminoea taylorae Petuch, 1987

Bivalvia
Mytilidae
Arcuatula papyria (Conrad, 1846)
Modiolus americanus (Leach, 1815)
Pectinidae
Argopecten irradians taylorae Petuch, 1987
Pinnidae
Atrina rigida (Lightfoot, 1786)
Atrina seminuda (Lamarck, 1786)
Atrina serrata (Sowerby I, 1825)

Carditidae
Carditamera floridana Conrad, 1838

Lucinidae
Divaricella dentata (Wood, 1815)
Phacoides pectinata (Gmelin, 1791)
Radiolucina amianta (Dall, 1901)
Stewartia floridana (Conrad, 1833)
Anodontia alba Link, 1807
Cardiidae
Laevicardium mortoni (Conrad, 1830)
Laevicardium serratum (Linnaeus, 1758)
Veneridae
Anomalocardia cuneimeris (Conrad, 1846)
Chione elevata (Say, 1822)
Semelidae
Semele proficua (Pulteney, 1799)
Tellinidae
Angulus sybariticus (Dall, 1881)
Solenocurtidae
Tagelus divisus (Spengler, 1794)

Figure 5.5 Gastropods of the Ten Thousand Islands Sea Grass Beds.
A, B= *Modulus floridanus* (Conrad, 1869), length 9 mm; **C, D**= *Crepidula ustulatulina* (Collin, 2002), length 8 mm; **E**= *Bulla occidentalis* A. Adams, 1850, length 27 mm; **F**= *Costoanachis semiplicata* (Stearns, 1873), length 14 mm; **G**= *Cerithium muscarum* Say, 1822, length 21 mm; **H**= *Haminoea taylorae* Petuch, 1987, length 9 mm; **I, J**= *Falsuszafrona taylorae* (Petuch, 1987), length 11 mm.

Figure 5.6 Muricid Gastropods of the Ten Thousand Islands Sea Grass Beds.
A, B= *Vokesimurex rubidus* (F.C. Baker, 1897), length 24 mm; **C, D=** *Phyllonotus pomum* (Gmelin, 1791), length 83 mm; **E, F=** *Chicoreus dilectus* (A. Adams, 1855), length 51 mm; **G, H=** *Chicoreus dilectus* (A. Adams, 1855), dark color form, length 56 mm.

Figure 5.7 The Western Florida Pear Whelk, *Fulguropsis pyruloides* (Say, 1822).
A, B= typical specimen, length 76 mm; **C, D=** orange color form, length 62 mm; **E, F=** pale specimen, length 78 mm.

Figure 5.8 The Florida Keys Pear Whelk, *Fulguropsis keysensis* Petuch, 2013.
A, B= orange color form, length 85 mm; **C, D**= typical specimen, length 87 mm;
E, F= dark color form, length 111 mm.

Figure 5.9 The Atlantic Alphabet Cone Shell, *Lindaconus atlanticus* (Clench, 1942). **A, B=** typical specimen, length 50 mm; **C, D=** exceptionally dark color form, length 48 mm; **E, F=** specimen with freak spire whorls, length 48 mm; **G=** *Lindaconus atlanticus* (color form *aureofasciatus* (Rehder & Abbott, 1951), labial view), length 54 mm.

Figure 5.10 The Florida Cone Shell, *Gradiconus floridanus* (Gabb, 1869).
A, B= typical specimen, length 41 mm; **C, D**= typical specimen, length 42 mm; **E, F**= form with a low spire, length 42 mm; **G, H**= typical specimen, length 39 mm. This well-known western Florida cone shell has erroneously been referred to as "*Gradiconus anabathrum* (Crosse, 1865)" by many recent cone workers. That name is now known to be a synonym of *Gradiconus scalaris* (Valenciennes, 1832), a species found only in the Eastern Pacific Panamic Province, and therefore Gabb's *floridanus* should be considered a valid and available name (see Berschauer, 2022).

Figure 5.11 The Florida Cone Shell, *Gradiconus floridanus* (Gabb, 1869) color form *floridensis* (Sowerby II, 1870).

A, B= typical specimen, length 37 mm; **C, D**= dark-colored specimen, length 47 mm; **E, F**= color form with rows of small dots, length 45 mm; **G, H**= exceptionally dark color form found on Camp Lulu Key, length 42 mm. This brightly-colored variety occurs in almost every population of the Florida Cone, but is most common in the central area of the Outer Ten Thousand Islands.

Figure 5.12 Bivalves of the Ten Thousand Islands Sea Grass Beds.
A= *Arcuatula papyria* (Conrad, 1846), length 24 mm; **B**= *Arcuatula papyria* (Conrad, 1864), length 21 mm; **C, D**= *Laevicardium mortoni* (Conrad, 1830), length 12 mm; **E**= *Stewartia floridana* (Conrad, 1833), length 26 mm; **F**= *Anomalocardia cuneimeris* (Conrad, 1846), length 15 mm; **G**= *Phacoides pectinata* (Gmelin, 1791), length 40 mm; **H**= *Radiolucina amianta* (Dall, 1901), length 9 mm.

Figure 5.13 Bivalves of the Ten Thousand Islands Sea Grass beds.
A= *Semele proficua* (Pulteney, 1799), length 28 mm; **B**= *Divaricella dentata* (Wood, 1815), length 24 mm; **C**= *Angulus sybariticus* (Dall, 1881), length 18 mm; **D**= *Carditamera floridana* Conrad, 1838, length 27 mm; **E**= *Chione elevata* Say, 1822, length 31 mm; **F**= *Modiolus americanus* (Leach, 1815), length 33 mm; **G**= *Tagelus divisus* (Spengler, 1794), length 23 mm.

Figure 5.14 Taylor's Bay Scallop, *Argopecten irradians taylorae* Petuch, 1987.
A, B= top and bottom valves of the typical color form, length 34 mm; **C, D**= top and bottom valves of an orange color form, length 36 mm; **E**= top valve of a large specimen of the typical color form, length 57 mm; **F**= top valve of a rare bright red color form, length 26 mm.

Figure 5.15 Pen Shells of the Ten Thousand Islands Sea Grass beds.
A= *Atrina rigida* (Lightfoot, 1786), length 268 mm; **B=** *Atrina seminuda* (Lamarck, 1786), length 148 mm; **C=** *Atrina serrata* (Sowerby I, 1825), length 149 mm.

CHAPTER 6. The Deep Channels and Offshore Areas

Probably the least-known and least-explored areas within the Ten Thousand Islands region are the deeper sea floors within the inter-island channels and offshore, outward into the South Florida Bight (Figures 6.1 and 6.2). Because of the tannin-stained water and suspended fine sediments, both derived from the effluent of Everglades rivers, little sunlight penetrates below a depth of 10 m, producing a permanent twilight on the deeper channel sea floors. With such reduced light conditions, sea grasses and algae cannot grow and there is a reduced primary productivity in these areas, leading to faunal impoverishment and a lowered biodiversity. Since they cannot support extensive plant growth, these deep, dark, and sunless areas often house large bioherms composed almost entirely of sponges and colonial tunicates (urochordata). Large flower-shaped masses of the delicate bryozoan ectoproct, *Schizoporella floridana* Osburn, 1914 (Figure 6.3), also grow in these areas, often covering large expanses of the muddy sea floors. The sponges, tunicates, and bryozoans regularly wash ashore on to the islands during heavy storms, and can often be found in large heaps along the high tide line. Because of the general murkiness of the water and low visibility, diving to the sea floor of the channels would yield little information and could prove dangerous, especially due to strong tidal currents and the abundant presence of Bull Sharks.

What little is known about the biodiversity of the deep channels comes from the activity of the commercial crab fishermen, who set their traps between the islands and in deeper offshore areas (5-25 m depths). Besides Blue and Stone Crabs, the baited traps (Figure 6.4) attract a host of carnivorous and scavenging gastropods, including large deep water specimens of the Tulip Shell, *Fasciolaria tulipa*, and the scaphelline Juno's Volute, *Scaphella junonia* (Lamarck, 1804) (Figures 6.9 and 6.10). This large spotted volute, the single most sought-after shell for collectors, normally hunts down and feeds upon the olive shell, *Americoliva sayana sarasotaensis*, but will also feed on freshly-dead fish and crabs. Many of the crab traps are also set in the deep water sponge beds and these attract several of the prominent resident gastropods, including the large Deer Cowrie, *Macrocypraea (Lorenzicypraea) cervus* (Linnaeus, 1771) (Figure 6.5), and the dwarf whelk *Hesperisternia multangula* (Philippi, 1848) (Figure 6.7). Deer Cowries are essentially omnivorous and will graze on algae, sponges, hydroids, ectoprocts, and carrion and often enter the traps in search of encrusting organisms and dead crabs. Of special interest on these deeper sea floors is a newly-discovered banded tulip shell, a distinct Ten Thousand Islands subspecies of the Tortugas Tulip Shell, which normally occurs in the Florida Keys, the Dry Tortugas, the Tortugas shrimping grounds, and the Florida Middle Grounds banks. This new endemic subspecies, *Cinctura tortugana foxi* Petuch and Berschuaer, 2022 (Figure 6.8) occasionally washes up on beaches, after heavy storms, on a stretch of coastline running from Kice and Marco Islands in the north to Pavilion Key in the south.

The open sand sea floors in the deep channels and offshore areas also house an interesting molluscan fauna that includes several widespread Carolinian Province gastropods, including the Milk Conch, *Macrostrombus costatus* (Gmelin, 1791) (Figure 6.6), the Fig Shell, *Ficus papyratia* (Say, 1822) (Figures 6.7A, B), the Nutmeg Shell, *Cancellaria reticulata* (Linnaeus, 1767) (Figures 6.7C, D), and the Scotch Bonnet, *Semicassis granulata* (Born, 1778) (Figures 6.7E, F, G), along with the widespread western Atlantic Spiny Murex, *Vokesimurex cabritii* (Bernardi, 1859). The small cassid *Semicassis granulata* feeds exclusively on sand dollars (irregular echinoid echinoderms), preferably the Five-Holed Sand Dollar, *Mellita quinquiesperforata* (Leske, 1778), which occurs in large shoals on deeper water sea floors. The sandy sea floors of the deeper channels and lagoons also house a number of interesting bivalves, including the Dwarf Razor Clam, *Ensis megistus* Pilsbry and McGinty, 1943 (Figures 6.11A, B, C), the Disc Venus, *Dosinia discus* (Reeve, 1850) (Figure 6.11D), the Ravenel's Surf Clam, *Spisula raveneli* (Conrad, 1832) (Figure 6.11E), and the delicate Keen's Lucine, *Callucina keenae* Chavan, 1971 (Figure 6.11F). Of special interest in the deeper open sand substrate is Vanhyning's Giant Cockle, *Dinocardium vanhyningi* Clench and Smith, 1944 (Figure 6.12), the largest member of its genus in the Gulf of Mexico and an endemic Gulf species that ranges from the Ten Thousand Islands around to Yucatan. After heavy storms, *Dinocardium vanhyningi* washed up on the beaches in great numbers and is one of the most conspicuous bivalves found in the Ten Thousand Islands.

Pictorial Overview of Deep Channels Areas and Associated Organisms

The following photographs show typical deep channel environments and some of their more conspicuous organisms that are present along the Ten Thousand Islands.

Figure 6.1 View of the deep channel in a branch of the Rabbit Key Pass, from a sand beach on Turtle Key, Ten Thousand Islands, Collier County. The large island across the pass is Lumber Key. Notice the coffee-brown color of the water caused by plant-derived tannins.

Figure 6.2 View of the mouth of the deep Chokoloskee Channel from an exposed section of the Worm Shell (*Petaloconchus nigricans*) reef platform on Turtle Key, Ten Thousand Islands, Collier County, looking northwestward toward Demijohn Key. A flock of Sanderlings, *Calidris alba* (Pallas, 1764), fattens-up on the exposed reef, feasting on small invertebrates before the annual migration to the Arctic.

Figure 6.3 Close-up view of a delicate, flower-like growth form (bryolith) of the bryozoan ectoproct, *Schizoporella floridana* Osburn, 1914, washed up onto the beach on Turtle Key, Ten Thousand Islands, Collier County, Florida. This type of ramose growth pattern is typical of *Schizoporella* bryoliths that grow on the sea floors of the deep inter-island channels, where they often form large biohermal structures. The flower-shaped bryolith washed up next to a fragment of a string of egg capsules of the Left-Handed Whelk and a bright rose-pink valve of the tellin *Eurytellina lineata*. Photo courtesy of Michael Bruggeman.

Figure 6.4 View of Stone Crab traps, stacked up for cleaning, along the Barron River in Everglades City, Collier County, Florida. The commercial Stone Crab fishermen bait these traps with chicken necks and dead fish and then drop them to depths of 10 m to 100 m within the deep inter-island channels or offshore in the South Florida Bight. The traps are left down for two weeks to a month and then retrieved after stone crabs have taken up residence. Besides crabs, the baited traps also attract large gastropods such as *Scaphella junonia* and *Macrocypraea (Lorenzicypraea) cervus*.

Iconography of Deep Channel and Offshore Mollusks

The gastropod and bivalve mollusks illustrated here are all classic inhabitants of the deep channels and areas offshore of the Ten Thousand Islands. The more important ecological index taxa include:

Gastropoda
Strombidae
Macrostrombus costatus (Gmelin, 1791)
Cypraeidae
Macrocypraea (Lorenzicypraea) cervus (Linnaeus, 1771)
Cassidae
Semicassis granulata (Born, 1778)
Ficidae
Ficus papyratia (Say, 1822)
Muricidae
Vokesimurex cabritii (Bernardi, 1859)
Fasciolariidae
Cinctura tortugana foxi Petuch and Berschauer, 2022
Pisaniidae
Hesperisternia multangula (Philippi, 1848)
Volutidae
Scaphella junonia (Lamarck, 1804)
Cancellariidae
Cancellaria reticulata (Linnaeus, 1767)

Bivalvia
Pectinidae
Argopecten gibbus (Linnaeus, 1758)
Euvola raveneli (Dall, 1898)
Lucinidae
Callucina keenae Chavan, 1971
Chamidae
Arcinella cornuta Conrad, 1866
Cardiidae
Dinocardium vanhyningi Clench and Smith, 1944
Veneridae
Dosinia discus (Reeve, 1850)
Pharidae
Ensis megistus Pilsbry and McGinty, 1943
Mactridae
Raeta plicatella (Lamarck, 1818)
Spisula raveneli (Conrad, 1832)

Figure 6.5 The Deer Cowrie, *Macrocypraea (Lorenzicypraea) cervus* (Linnaeus, 1771), the largest species of Cowrie Shell.

A, B= typical specimen, length 124 mm; **C, D=** specimen with a color pattern of small circles, length 128 mm. Small specimens of *Macrocypraea (Lorenzicypraea) cervus* are common on offshore sponge reefs.

Figure 6.6 The Milk Conch, *Macrostrombus costatus* (Gmelin, 1791).
A, B= typical specimen, length 143 mm.

Figure 6.7 Gastropods of Deeper Offshore Areas.
A, B= *Ficus papyratia* (Say, 1822), length 85 mm; **C, D=** *Cancellaria reticulata* (Linnaeus, 1767), length 32 mm; **E, F=** *Semicassis granulata* (Born, 1778), typical specimen, length 76 mm; **G, H=** *Hesperisternia multangula* (Philippi, 1848), typical color form, length 28 mm; **I=** *Hesperisternia multangula* (Philippi, 1848), orange color form, length 26 mm; **J=** *Vokesimurex cabritii* (Bernardi, 1859), length 45 mm, rarely washes up onto the beaches near Marco Island.

Figure 6.8 Fox's Tulip Shell, *Cinctura tortugana foxi* Petuch and Berschauer, 2022. **A, B**= holotype, length 79 mm; **C, D**= typical specimen, length 87 mm; **E**= typical specimen, length 71 mm; (**A-D** were collected on the beach at Kice Island, Collier County); **F**= variant with multiple color bands, length 81 mm, a faded beach specimen found on Camp Lulu Key after a heavy storm; **G, H**= *Cinctura tortugana* (Hollister, 1957), length 74 mm, from 60 m depth north of the Dry Tortugas, Florida Keys, Monroe County, Florida; for comparison with the new subspecies *tortugana foxi*.

Figure 6.9 The Juno's Volute, *Scaphella junonia* (Lamarck, 1804).
A, B= typical Ten Thousand Islands specimen, length 108 mm; **C, D**= typical specimen, length 110 mm. Fragments of *Scaphella junonia* have been found on Pavilion Key in the southern Ten Thousand Islands, and fresh specimens regularly wash onto the beach at Marco and Kice Islands in the Cape Romano area.

Figure 6.10 Color Variants of the Juno's Volute, *Scaphella junonia*.
A= specimen with the checkers having fused into longitudinal stripes, length 91 mm; **B, C**= specimen with fused spots, length 121 mm.

Figure 6.11 Bivalves of the Deep Channels and Offshore Areas.
A= *Ensis megistus* Pilsbry and McGinty, 1943, typical specimen, length 66 mm; **B**= *Dosinia discus* (Reeve, 1850), length 64 mm; **C**= *Spisula raveneli* (Conrad, 1832), length 50 mm; **D**= *Callucina keenae* Chavan, 1971, length 36 mm; **E**= *Raeta plicatella* (Lamarck, 1818), length 47 mm; **F**= *Argopecten gibbus* (Linnaeus, 1758), length 30 mm; **G**= *Euvola raveneli* (Dall, 1898), length 51 mm; **H**= *Arcinella cornuta* Conrad, 1866, length 34 mm; **I**= *Callista maculata* (Linnaeus, 1758, length 39 mm.

Figure 6.12 The Gulf Giant Cockle, *Dinocardium vanhyningi* Clench and Smith, 1944.
A= length 61 mm; **B, C**= length 135 mm, front and side views of the same specimen.

Figure 6.13 A flock of the American White Ibis, *Eudocimus albus* (Linnaeus, 1758), heads home to its mangrove forest rookery after a day of feeding on the mud flats in Chokoloskee Bay, Ten Thousand Islands, Collier County. A large thunderhead, reflecting the setting sun, forms the perfect backdrop for these iconic Everglades birds. Photo courtesy of Michael Bruggeman.

Systematic List of the Mollusks of the Ten Thousand Islands

The following species list is an updated, expanded, and emended version of the species list given by Petuch and Myers, 2014 (pp. 156-167). As can be seen from this faunal list, only 113 species and several named forms of mollusks, including both gastropods and bivalves, are now known from the Ten Thousand Islands. These were sampled over a period of time extending from 2013 to 2021. This small number contrasts greatly with more than 1,200 species of mollusks known from the nearby Florida Keys (Petuch and Myers, 2014). As discussed in the Introduction, this faunal impoverishment is due to the Ten Thousand Islands area having cooler water temperatures during the winter months and having wildly fluctuating salinities during the hot, rainy summer months. These harsh and ecologically-stressful conditions allow only the hardiest and most physiologically-plastic organisms to survive, resulting in a greatly impoverished malacofauna. The molluscan fauna of the Ten Thousand Islands is listed here systematically by family.

Gastropoda
Fissurellidae
Diodora listeri (d'Orbigny, 1842)
Lucapinella suffusa (Reeve, 1850)
Calliostomatidae
Calliostoma tampaense (Conrad, 1846)
Neritidae
Vitta usnea (Roding, 1798)
Littorinidae
Littoraria angulifera (Lamarck, 1822)
Littoraria irrorata sayi (Philippi, 1846)
Littoraria nebulosa (Lamarck, 1822)
Littoraria nebulosa form *tessellata* (Philippi, 1847)
Batillariidae
Lampanella minima (Gmelin, 1791)
Cerithiidae
Cerithium muscarum Say. 1822
Modulidae
Modulus floridanus (Conrad, 1869)
Strombidae
Macrostrombus costatus (Gmelin, 1791)
Strombus alatus Gmelin, 1791
Crepidulidae
Crepidula fornicata (Linnaeus, 1758)
Crepidula ustulatulina Collin, 2002
Ianacus atrasolea (Collin, 2002)
Ianacus plana (Say, 1822)
Vermetidae
Petaloconchus nigricans (Dall, 1884)
Turritellidae
Vermicularia fargoi owensi Petuch and Myers, 2014

Cypraeidae
Macrocypraea (Lorenzicypraea) cervus (Linnaeus, 1771)
Naticidae
Naticarius canrena (Linnaeus, 1758)
Naticarius verae Rehder, 1947
Neverita (Glossaulax) delessertiana (Recluz, 1843)
Neverita (Glossaulax) duplicata campechiensis (Recluz, 1843)
Cassidae
Semicassis granulata (Born, 1778)
Ficidae
Ficus papyratia (Say, 1822)
Muricidae
Chicoreus dilectus (A. Adams, 1855)
Eupleura tamapensis (Conrad, 1846)
Phyllonotus pomum (Gmelin, 1791)
Vokesimurex cabritii (Bernardi, 1859)
Vokesimurex rubidus (F.C. Baker, 1897)
Vokesinotus perrugatus (Conrad, 1846)
Fasciolariidae
Cinctura hunteria (Perry, 1811)
Cinctura tortugana foxi Petuch and Berschauer, 2022
Fasciolaria tulipa (Linnaeus, 1758)
Triplofusus giganteus (Kiener,1840)
Busyconidae
Fulguropsis keysensis Petuch, 2013
Fulguropsis pyruloides (Say, 1822)
Sinistrofulgur sinistrum (Hollister, 1958)
Nassariidae
Phrontis vibex (Say, 1822)
Uzita swearingeni Petuch and Myers, 2014
Pisaniidae
Gemophos tinctus pacei Petuch and Sargent, 2011
Hesperisternia multangula (Philippi, 1848)
Melongenidae
Melongena (Rexmela) corona (Gmelin, 1791)
Melongena (Rexmela) corona (Gmelin, 1791) form *belknapi* Petit de la Saussaye, 1850
Melongena (Rexmela) corona (Gmelin, 1791) form *perspinosa* Pilsbry and Vanatta, 1934
Melongena (Rexmela) corona (Gmelin, 1791) form *trinodosa* Emery and Lermond, 1936
Columbellidae
Costanachis semiplicata (Stearns, 1873)
Falsuszafrona taylorae (Petuch, 1987)
Olividae
Americoliva sayana sarasotaensis Petuch and Sargent, 1986
Volutidae
Scaphella junonia (Lamarck, 1804)
Marginellidae
Prunum apicinum (Menke, 1828)
Cancellariidae
Cancellaria reticulata (Linnaeus, 1758)

Conidae
Gradiconus floridanus (Gabb, 1868)
Gradiconus floridanus form *floridensis* (Sowerby II, 1870)
Jaspidiconus acutimarginatus (Sowerby II, 1866)
Jaspidiconus stearnsi (Conrad, 1869)
Lindaconus atlanticus (Clench, 1942)
Lindaconus atlanticus form *aureofasciatus* (Rehder and Abbott, 1951)
Terebridae
Neoterebra dislocata (Say, 1822)
Neoterebra vinosa (Dall, 1889)
Bullidae
Bulla occidentalis A. Adams, 1850
Melampidae
Melampus bidentatus (Say, 1822)
Melampus coffea (Linnaeus, 1758)
Melampus monilis (Bruguiere, 1789)
Ellobiidae
Ellobium (Auriculoides) dominicense (Ferussac, 1821)
Orthalicidae (Terrestrial-Arboreal)
Liguus fasciatus castaneozonatus form *mitchelli* Poland, 2008
Orthaulax floridensis (Pilsbry, 1899)
Bivalvia
Arcidae
Anadara floridana (Conrad, 1869)
Anadara transversa (Say, 1822)
Noetiidae
Noetia ponderosa (Say, 1822)
Mytilidae
Arcuatula papyria (Conrad, 1846)
Geukensia granosissima (Sowerby III, 1914)
Ischadium recurvum (Rafinesque, 1820)
Modiolus americanus (Leach, 1815)
Ostreidae
Crassostrea rhizophorae (Guilding, 1828)
Crassostrea virginica (Gmelin, 1791)
Ostreola equestris (Say, 1834)
Pinnidae
Atrina rigida (Lightfoot, 1786)
Atrina seminuda (Lamarck, 1786)
Atrina serrata (Sowerby I, 1825)
Pectinidae
Argopecten gibbus (Linnaeus, 1758)
Argopecten irradians taylorae Petuch, 1987
Euvola raveneli (Dall 1898)
Anomiidae
Anomia simplex d'Orbigny, 1853
Carditidae
Carditamera floridana Conrad, 1838

Lucinidae
Anodontia alba Link, 1807
Callucina keenae Chavan, 1971
Divaricella dentata (Wood, 1815)
Phacoides pectinata (Gmelin, 1791)
Radiolucina amianta (Dall, 1901)
Stewartia floridana (Conrad, 1833)

Chamidae
Arcinella cornuta Conrad, 1866

Cardiidae
Dallocardia muricata (Linnaeus, 1758)
Dinocardium vanhyningi Clench and Smith, 1944
Laevicardium mortoni (Conrad, 1830)
Laevicardium serratum (Linnaeus, 1758)
Trachycardium egmontianum (Shuttleworth, 1856)

Veneridae
Anomalocardia cumeimeris Conrad, 1846)
Callista maculata (Linnaeus, 1758)
Chione elevata Say, 1822
Dosinia discus (Reeve, 1850)
Macrocallista nimbosa (Lightfoot, 1786)
Mercenaria browni Petuch and Berschauer, 2019
Mercenaria campechiensis (Gmelin, 1791)
Mercenaria mercenaria notata (Say, 1822) (an endemic North Carolina species which was introduced on Cedar Key to establish a commercial clamming industry; the clam is now invasive along western Florida)

Tellinidae
Angulus sybariticus (Dall, 1881)
Eurytellina lineata (Turton, 1819)
Macoma tenta (Say, 1834)
Tampaella tampaensis (Conrad, 1866)

Semelidae
Semele proficua (Pulteney, 1799)

Solenocurtidae
Tagelus divisus (Spengler, 1794)

Pharidae
Ensis megistus Pilsbry and McGinty, 1943

Mactridae
Mactrotoma fragilis (Gmelin, 1791)
Raeta plicatella (Lamarck, 1818)
Spisula raveneli (Conrad, 1832)

Pholadidae
Barnea truncata (Say, 1822)
Cyrtopleura costata (Linnaeus, 1758)

Periplomatidae
Periploma margaritaceum (Lamarck, 1801)

REFERENCES

Berschauer, D.P. 2022. The True Identity of *Gradiconus anabathrum* (Crosse, 1865). *Xenophora Taxonomy* 35:43-48.

Close, H.T. 2000. *The Liguus Tree Snails of Southern Florida*. University Press of Florida, Gainseville, Florida. 161 pp.

Gause, G.F. 1934. *The Struggle for Existence (First Edition).* Williams & Wilkins Publishers, Baltimore. 320 pp.

González-Guillén, A., F. Krull, L.A. Lajonchere-Ponce de León. 2018. *Liguus The flamboyant tree snails.* InstantPublishers.com, Memphis, Tennessee. 498 pp.

Macmahon, D.A. and W.H. Marquardt. 2004. *The Calusa and Their Legacy: South Florida People and Their Environments.* University Press of Florida, Gainesville, Florida. 240 pp.

Marquardt, W.H. and L. Kozuch. 2016. The lightning whelk: An enduring icon of southeastern North American spirituality. *Journal of Anthropological Archaeology* 42:1-26.

Petuch, E.J. 2013. *Biogeography and Biodiversity of Western Atlantic Mollusks*. CRC Press, London, New York, Boca Raton. 234 pp.

Petuch, E.J. and D.P. Berschauer. 2019. New Species of Mollusks (Gastropoda and Bivalvia) from the Tropical Western Atlantic, West Africa, and Red Sea. *The Festivus* 51(3):218-230.

Petuch, E.J. and D.P. Berschauer. 2020a. A Review of the *Cinctura* Banded Tulips (Gastropoda: Fasciolariidae), with the Descriptions of Four New Subspecies and a New Subgenus. *The Festivus* 52(4):316-334.

Petuch, E.J. and D.P. Berschauer. 2020b. *Tropical Marine Mollusks: An Illustrated Biogeographical Guide.* CRC Press, London, New York, Boca Raton. 357 pp.

Petuch, E.J. and D.P. Berschauer. 2021. *Ancient Seas of Southern Florida: The Geology and Paleontology of the Everglades Region*. CRC Press, London, New York, Boca Raton. 284 pp.

Petuch, E.J. and R.F. Myers. 2014. *Molluscan Communities of the Florida Keys and Adjacent Areas*. CRC Press, London, New York, Boca Raton. 300 pp.

Petuch, E.J. and C.E. Roberts. 2007. *The Geology of the Everglades and Adjacent Areas*. CRC Press, London, New York, Boca Raton. 212 pp.

Petuch, E.J. and D.M. Sargent. 2011. *Rare and Unusual Shells of Southern Florida (Mainland, Florida Keys, Dry Tortugas)*. Conch Republic Books, Mount Dora, Florida. 187 pp.

Poland, P.I. 2008. *Liguus fasciatus* var. *mitchelli* (proposed). The *Liguus* Discussion Board, Dec. 25, 2008. https://www.tapatalk.com/groups/liguusdiscussionboard/mitchelli-t1082.html

Shier, D.E. 1969. Vermetid Reefs and coastal development in the Ten Thousand Islands, southwest Florida. *Geological Society of America Bulletin* 80:485-508.

Thompson, V.D., W.H. Marquardt, M. Savarese, K.J. Walker, L.A. Newsom, I. Lulewicz, N.R. Lawres, A.D. Roberts Thompson, A.R. Bacon, and C.A. Walser. 2020. Ancient engineering of fish capture and storage in southwest Florida. Proceedings of the National Academy of Sciences 117(15):8374-8381. DOI: 10.1073/pnas.1921708117

Thompson, V.D., W.H. Marquardt, A. Cherkinsky, A.D. Roberts Thompson, K.J. Walker, L.A. Newsom, and M. Savarese. 2016. From Shell Midden to Midden-Mound: The Geoarchaeology of Mound Key, an Anthropogenic Island in Southwest Florida, USA. *PLoS ONE* 11(4):e0154611. https://doi.org/10.1371/journal.pone.0154611

White, W.A. 1970. *Geomorphology of the Florida Peninsula*. Florida Geological Survey Bulletin 51. Tallahassee, Florida. 185 pp.

View of the lagoon on Jewell Key.

INDEX

The following alphabetical listing includes marine ecological terms, geographical localities within the Ten Thousand Islands region, and non-molluscan organisms that are illustrated or discussed in the book. Molluscan taxa are not listed here, as there is a systematically-arranged list of the Ten Thousand Islands malacofauna in a previous section.

View of the beach on Lumber Key.

ABOUT THE AUTHORS

Edward J. Petuch was born in Bethesda, Maryland, in 1949. Raised in a Navy family, he spent many of his childhood years collecting living and fossil shells in such varied localities as Chesapeake Bay, California, Puerto Rico, and Wisconsin. His early interests in malacology and oceanography eventually led to BA and MS degrees in zoology from the University of Wisconsin-Milwaukee. During his MS thesis research, Petuch concentrated on the molluscan biogeography of West Africa, traveling extensively in the Canary Islands, Western Sahara, Senegal, Gambia, Sierra Leone, Ivory Coast, and Cameroons. During this time, he also conducted research on the molluscan ecology of both coasts of Mexico and the Great Barrier Reef of Belize.

Continuing his education, Petuch studied marine biogeography and malacology under Gilbert Voss and Donald Moore at the Rosenstiel School of Marine and Atmospheric Sciences at the University of Miami, where he received a full scholarship. During this time, his doctoral dissertation research involved intensive field work in Costa Rica, Colombia, Venezuela, Barbados, the Grenadines, and Brazil, where he often went to sea with the local shrimpers for weeks at a time. After receiving his PhD in oceanography in 1980, Petuch was invited to conduct two years of postdoctoral research, funded by the National Science Foundation, with Geerat Vermeij at the University of Maryland. While there, he also held a research associateship with the Department of Paleobiology at the National Museum of Natural History, Smithsonian Institution, under the sponsorship of Thomas Waller, and conducted field work in the Plio-Pleistocene fossil beds of Florida and North Carolina and the Miocene fossil beds of Maryland and Virginia.

Petuch has also collected and studied living mollusks in Australia, Papua-New Guinea, Fijis, French Polynesia, Japan, the Bahamas, Nicaragua, and Uruguay. This research has led to the publication of over 350 scientific papers and the discovery and description of almost 2,000 new species of mollusks and almost 200 new genera. His previous 23 books are well-known reference texts in the malacological and paleontological communities, and some of the better known include: *Ancient Seas of Southern Florida: The Geology and Paleontology of the Everglades Region* (2021), *Tropical Marine Mollusks: An Illustrated Biogeographical Guide* (2020), *Jewels of the Everglades: The Fossil Cowries of Southern Florida* (2018), *The Living and Fossil Busycon Whelks: Iconic Mollusks of Eastern North America* (2015), *Cone Shells of the Okeechobean Sea* (2015), *Molluscan Communities of the Florida Keys and Adjacent Areas: Their Ecology and Biodiversity* (2014), *Biogeography and Biodiversity of Western Atlantic Mollusks* (2013), *Molluscan*

Paleontology of the Chesapeake Miocene (2010), *The Geology of the Everglades and Adjacent Areas* (2007), and *Cenozoic Seas: The View from Eastern North America* (2004). Currently, Petuch is a Professor Emeritus in the Department of Geosciences, Florida Atlantic University in Boca Raton, Florida where, for thirty years, he taught undergraduate classes in oceanography, paleontology, and physical geology, and graduate classes in paleoecology and paleoceanography. He currently resides in Jupiter, Florida, with his wife Linda, where they both enjoy visits from their three children and their families.

© Morgan Taylor Photography, with permission

David P. Berschauer was born in Rockville Center, New York, in 1964, and spent his youth collecting shells in such varied localities as New York, Florida, California, Washington, Mexico, and throughout the Caribbean. His early interests in natural history, malacology, and marine biology eventually led to a BS in biology at the University of California-Irvine, an advanced marine invertebrate zoology course at Washington State University's Friday Harbor Marine Lab, and studies towards the pursuit of a graduate degree in marine biology at Florida State University in Tallahassee, Florida. While still a college undergraduate, Berschauer performed field biology research, published a number of research papers and gave scientific presentations at national conferences. He subsequently switched career paths and attended Southwestern University School of Law in Los Angeles, California, earning his Juris Doctorate in 1991. Although having developed a legal career, he has kept malacology as a lifetime avocation and has put together a sizeable research collection and personal museum of molluscan specimens.

Over his entire professional life, Berschauer continued to pursue his passion for marine biology, and collecting and studying marine organisms. In his spare time, he has developed and published a relational database software program to aid in the organization and maintenance of a systematic collection. Although originally designed for malacology, the program is applicable to entomology and other aspects of systematic zoology. Berschauer is an active member of the San Diego Shell Club, is a Museum Associate with the Natural History Museum of Los Angeles County ("LACM") Malacology Department, and is the Editor-in-Chief of the journal, *The Festivus*. He is also well known for his natural history and shell photography, with a multitude of high-quality examples seen throughout this book.

Besides being the author of many important scientific papers on molluscan systematics, Berschauer has described and named over 100 new species of gastropods and is also the co-author of several recent books on mollusks, including *The Living and Fossil Busycon Whelks: Iconic Mollusks of Eastern North America* (2015), *Jewels of the Everglades: The Fossil Cowries of Southern Florida* (2018), the well-received *Sea Shells of Southern California: Marine Shells of the Californian Province* (2018), *Tropical Marine Mollusks: An Illustrated Biogeographical Guide* (2020), and *Ancient Seas of Southern Florida: The Geology and Paleontology of the Everglades Region* (2021). He currently resides in Cairo, Georgia, with his wife Felicia, where they both enjoy visits with their daughter, son-in-law, and new grandson Lincoln.

Capt. Craig Daniels (left), field assistant Patrick Wilson (middle), and the senior author (right), being rescued after becoming stranded on Pavilion Key during a freak wind storm.

APPENDIX 1. Map of the Ten Thousand Islands: Northern Section

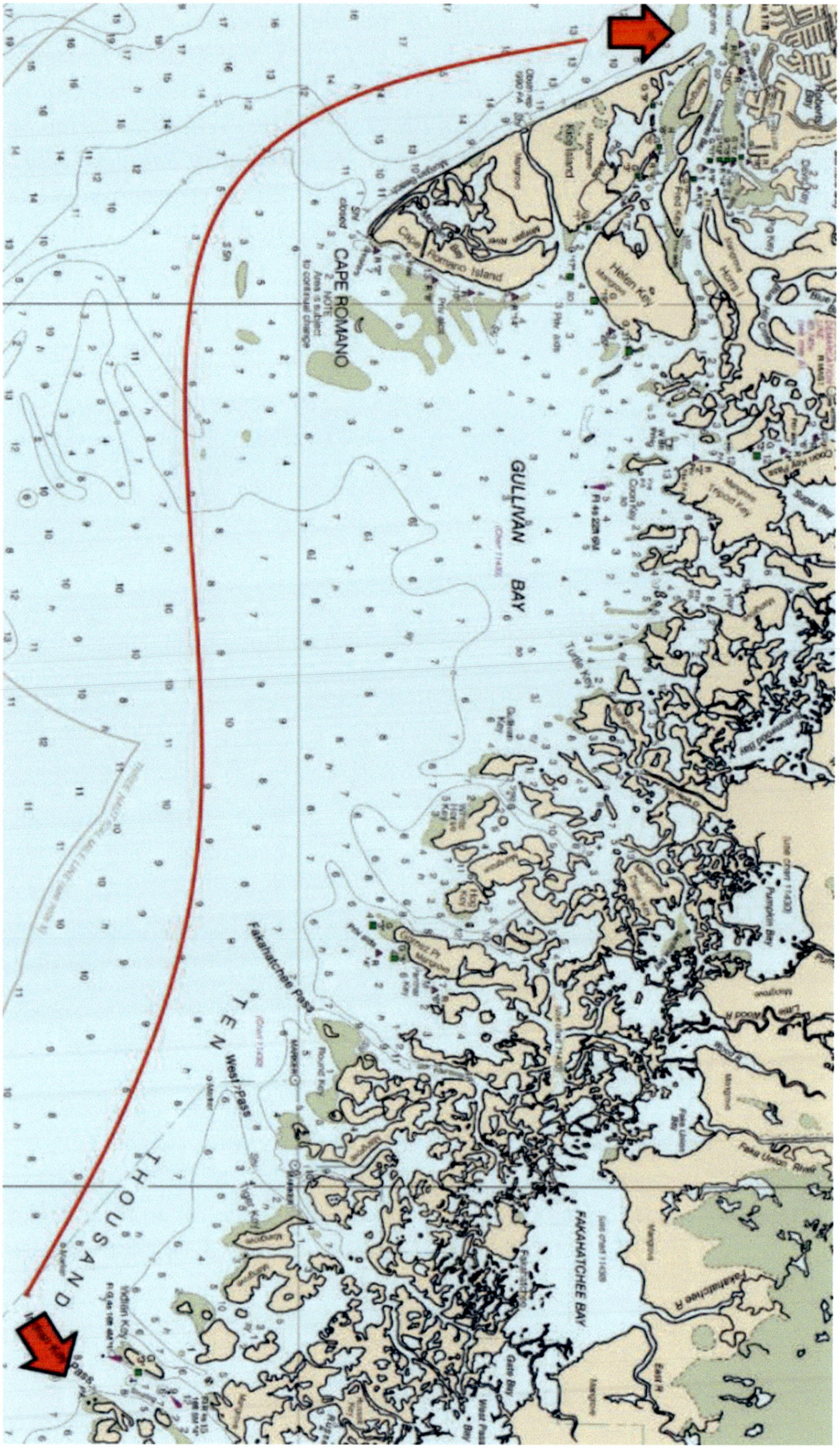

Map 1. Marco and Cape Romano Island to Indian Key Pass

APPENDIX 2. Map of the Ten Thousand Islands: Southern Section (upper half)

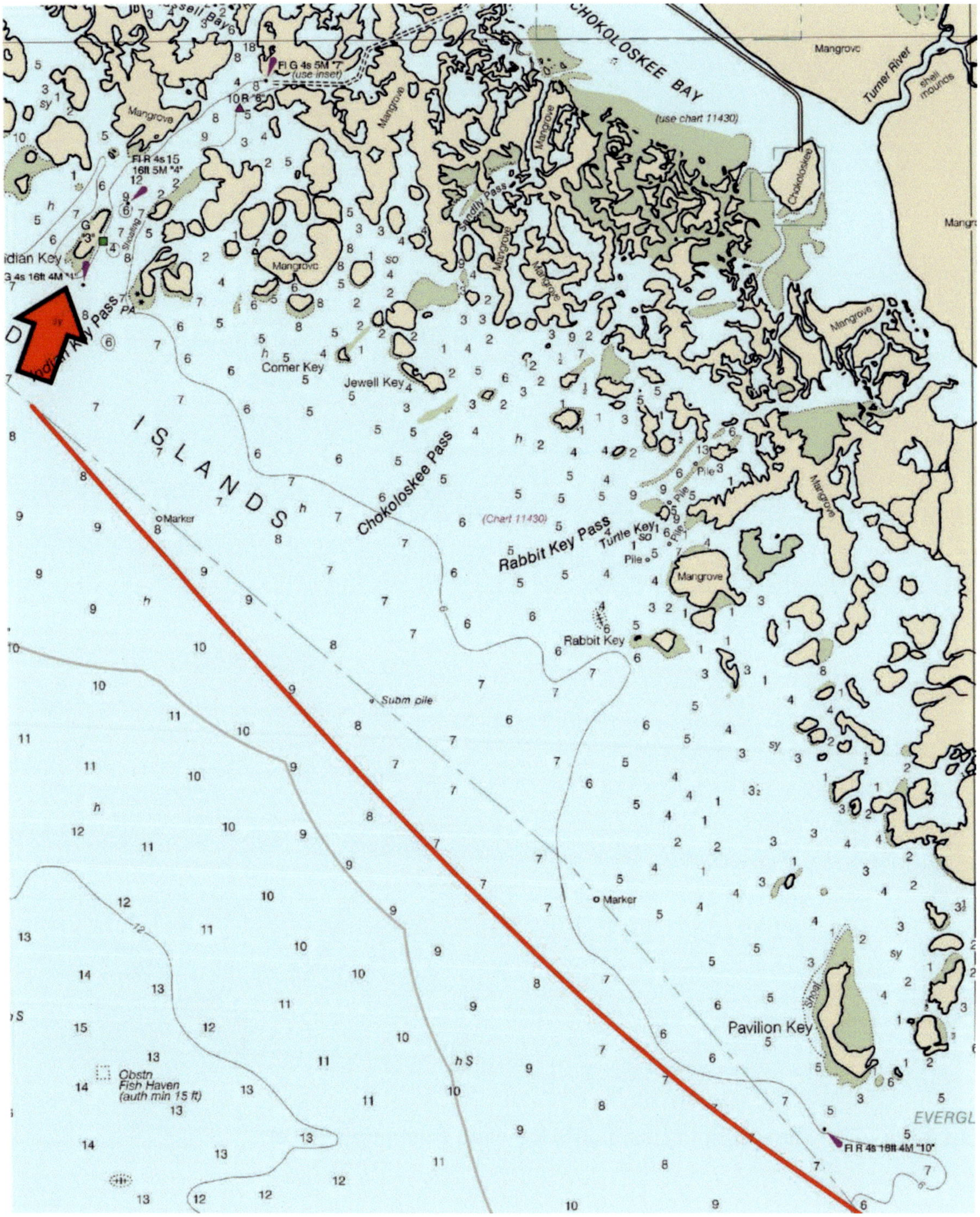

Map 2. Indian Key Pass south to Pavilion Key

APPENDIX 3. Map of the Ten Thousand Islands: Southern Section (lower half)

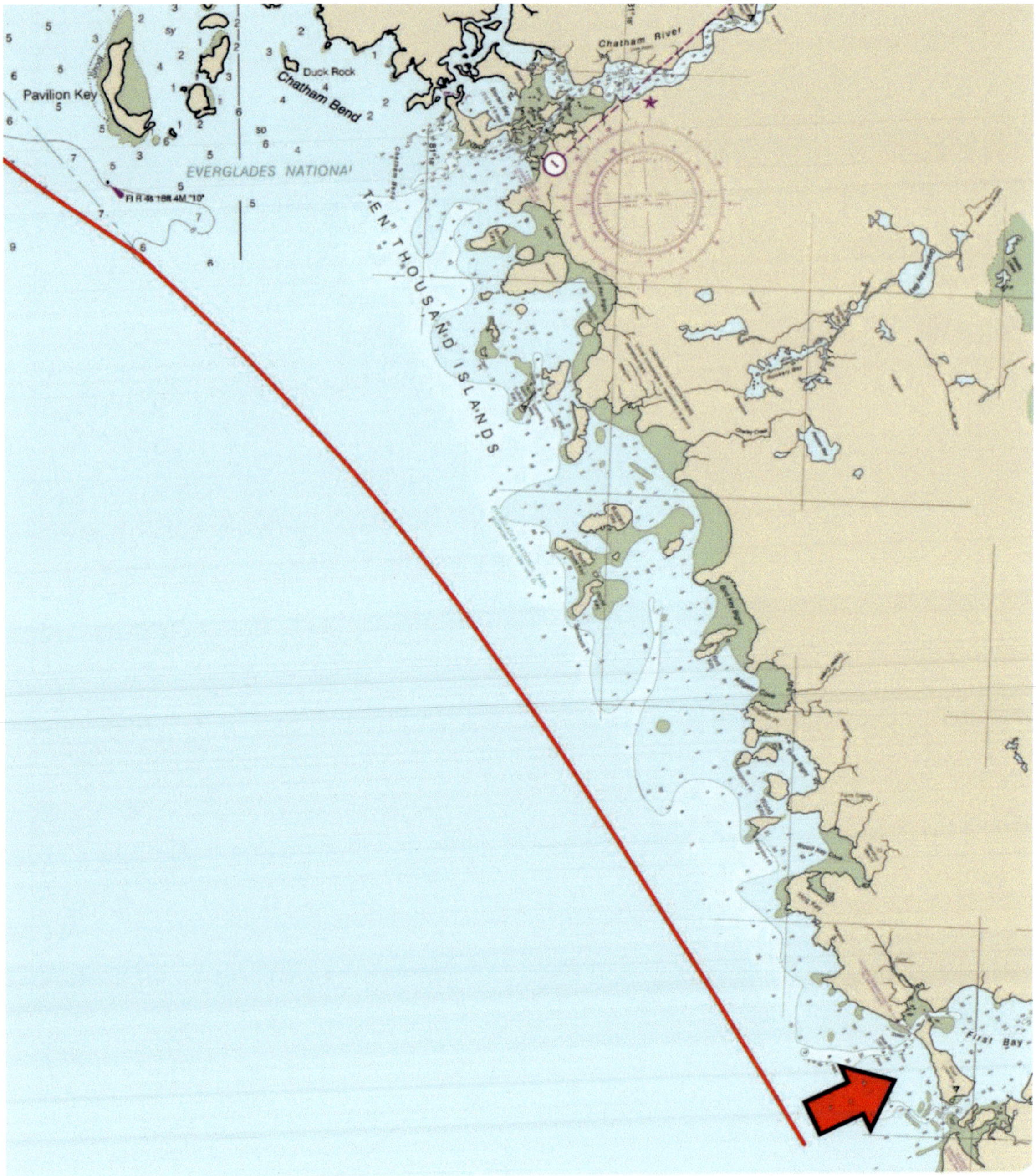

Map 3. Pavilion Key south to Lostman's Key and Lostman's River